MAKING OPEN INNOVATION WORK

@lindegaard to big and small companies:
You need to open up your innovation efforts!
Read this book and visit www.15inno.com for good advice.

STEFAN LINDEGAARD

Copyright 2011 STEFAN LINDEGAARD
All rights reserved.
ISBN: 1463712448
ISBN 13: 9781463712440
Library of Congress Control Number: 2011912053
CreateSpace, North Charleston, SC

TABLE OF CONTENTS

Introduction: From Why to How   v

I: Definitions   1

II: The Benefits of Open Innovation   13

III: When Big Companies Meet Small Companies In the Open Innovation Efforts   29

IV: Getting Your Organization Ready for Open Innovation   45

V: Getting Your People Ready for Open Innovation   63

VI: What to Consider Before Leaping into a Partnership   71

VII: Making It Work   93

VIII: Why Things Go Wrong   105

IX: Innovation Marketplaces   113

X: IPR and Open Innovation   125

XI: Using Social Media Tools	135
XII: Key Chapter Takeaways Recap	149
Appendix A The Ten Types of Innovation™	161

INTRODUCTION

FROM WHY TO HOW

Over the past few years, many of the leading global companies have begun to embrace open innovation. They reach outside their organizations to partner with suppliers, customers, academics, competitors, and entrepreneurs in the search for great new ideas that will keep their businesses on the cutting edge. Much has been written about this shift from a closed innovation model, in which all aspects of the innovation process take place internally in a tightly controlled, highly secretive environment, to an open model in which external partners play a key role in helping a company innovate. My first book, *The Open Innovation Revolution: Essentials, Roadblocks, and Leadership Skills* examined this shift.

Relatively little attention has been paid, however, to the way big and small companies intersect in open innovation. The assumption seems to be that open innovation works the same for everybody, but, in truth, it does not. Small companies have different resources, different needs, and different ways of contributing to an open innovation relationship.

Viewing open innovation for small companies the same way we view it for big companies is riddled with problems and has the potential to set small companies up for failure. This book explains why open innovation is different in small companies than it is in big companies and how small as well as big companies need to approach working together. My primary focus will be on how small and big companies can benefit from pursuing open innovation *together* and how they can make the intersection of their two organizations work.

What Is a Small Company?

For the purposes of this book, I define small companies as those that have only one key product line. They may have many employees, but all those workers are focused on developing, producing, and selling just that one product line. If they don't succeed at that, the company fails; they either have no Plan B or they are only in the early stages of developing it.

Also, the small companies I'm talking about do not have an innovation officer, someone who is in charge of making innovation happen; the chief executive officer is the one who drives innovation. In fact, they often don't even have an innovation strategy. They just try to improve that one product line without thinking much about how they could grow the company by adding other platforms. For the small companies that do begin to add more product lines, then the need for a chief innovation officer and an innovation strategy begins to become apparent.

Many of these small companies don't have a clue how big companies view open innovation. You will notice this in some of the interviews with smaller companies in the book. So what happens when such a company intersects with a large company that is interested in pulling them into their open innovation ecosystem? If the small company doesn't get educated fast, they could be at a severe disadvantage. Not all big companies play nice, so the smaller company could end up with the short end of the deal. In the long run, both small and big companies will lose if they do not meet on equal terms.

Some people argue that small companies are not big enough for open innovation because they don't have the organizational infrastructure to engage in open innovation. Other people do believe small companies have a role to play in open innovation ecosystems, but believe they inevitably will get the backseat while the big companies take the driver's seat. While I disagree with the first statement, I agree with the second. In open innovation, companies either control the projects or they contribute to them. Big companies prefer to be in control of their projects, whereas smaller companies do not even get a choice unless they have something unique to offer, which is rarely the case.

I believe small companies have an important role to play in open innovation, both for themselves and for large corporations. For small companies, an open innovation partnership can provide access to resources needed to hit it big, such as distribution channels or production resources. In many cases, small companies have been founded by entrepreneurs who have tremendous scientific and technological expertise. Corporations need access to this expertise and the innovative ideas and solutions that can grow out of it.

I like how John Conoley, CEO of Psion, pioneer of quality mobile handheld computers and their application in industrial markets around the world, explained why his company needs to develop a strong open innovation process and partnerships: "We decided to embrace open innovation to be faster and competitively unpredictable."

What company wouldn't want to be quicker to market with surprising new products and services? That's what open innovation can help you achieve, no matter your company's size. This book thoroughly

Introduction

explores the open innovation intersection between small and big companies, looking at topics that include:

- What is open innovation? What forms does it take and what benefits does it provide for both large and small companies, and what challenges does it pose for companies of different sizes?
- Case studies illustrating open innovation at work in small companies
- How to identify and develop the people who will drive open innovation in your organization
- Ways to develop open innovation capabilities within a small company
- Why big companies need small companies as part of their open innovation ecosystems
- Strategies for building and making open innovation partnerships work when the partners are of unequal size
- What to do if things go awry in an open innovation partnership
- How to handle issues of intellectual property (IP) rights
- How to use social media tools to build your open innovation capabilities and attract partnerships
- Throughout this book, you will hear from other open innovation experts and from people in companies who are charged with making open innovation happen. Each chapter ends with a list of key takeaways: these are gathered in one place in the book's final chapter so you have one easy

place to go whenever you want to quickly review what you've learned.

This Is the Future

Here's something very important that I want to emphasize before we go any further. In five to seven years, we will no longer talk about open innovation. The term "open innovation" will disappear and we will just view this as "innovation." The key difference is that innovation will have a much higher external input that what we see today.

This evolution will take longer in some industries than in others, but ultimately everyone will get to the same place—using an open innovation model. Thus, there is no need to set up an entirely different open innovation unit; rather, you should build open innovation capabilities in as part of your company's innovation DNA.

Open innovation is evolving so rapidly that I encourage you to visit my blog at www.15inno.com to find new case studies and new ideas about this important topic. Here you can also respond to my posts and share your experiences with open innovation. Let's keep the conversation going!

I
DEFINITIONS

Enough people still ask me to define open innovation that it's clear to me that plenty of confusion still exists as to what we're talking about when we refer to this topic. This may be particularly true at small companies since many businesses in this market segment are only just now beginning to test the waters of open innovation. So let's make sure we're all on the same page about how to define open innovation and related terms such as crowdsourcing, user-driven innovation, ecosystems, and innovation intermediaries.

When people ask what open innovation is, I suggest viewing open innovation as a philosophy or a mindset that you should embrace within you organization. In a more practical definition, open innovation is about bridging internal and external resources and acting on those opportunities to make innovation happen. I also like this quote from Henry Chesbrough: "Open innovation is a paradigm that assumes that firms can and should use external ideas as well as internal ideas, and internal and external paths to market, as the firms look to advance their technology."[1]

This is in contrast to the old model of closed innovation, in which a company maintained complete control over all aspects of the innovation process and discoveries were kept highly secret. In closed innovation, you do not attempt to assimilate input from outside sources into the innovation process, and you avoid having to share intellectual property or profits with any outside source. If this sounds familiar to small companies, that's probably because it may well be the model your company has used to develop its product offerings to date.

Also, in a closed innovation environment, activities are often segregated within an R&D department where the best and the brightest are expected to make sure the company gets to market early with new ideas to gain the "first mover" advantage.

In contrast, in open innovation your company works with external companies during the innovation process. Open innovation is often about soliciting ideas from outside, but it goes deeper than just involving others in the idea generation phases; the contribution from outside your company must be significant. It is also more than just a partnership where you pay for specific services. Everyone involved in an open innovation process focuses on problems, needs, and issues and works them out *together*. Furthermore, you can argue that closed innovation primarily focuses on products and services, whereas you are more likely to use open innovation to work with a broader range of the Ten Types of Innovation (see Appendix A), including business models, channels, and processes.

An important point to note: too often, companies fall into the trap of considering open innovation approaches only during what is called the front end of innovation. Once they have gotten input from external sources during the front end of innovation, they do everything themselves. Granted, it is a good thing to get more diverse input early on, but why stop there?

Open innovation is an approach that can work and should be used in any stage of the innovation process, not just in the early phases during the front end of innovation.

Companies miss out on the full potential of open innovation when they shut down to external resources later in the process. Perhaps companies are more willing to accept external input during the front end of innovation, when everyone expects more creativity and openness. Once this phase is completed, companies go into execution mode, which is less complex if they only use the internal resources they already know very well.

Related Terms

Where do crowdsourcing, user-driven innovation and co-creation fit into the open innovation picture? In general, these are tools and techniques that can be used within the paradigm of open innovation, but these terms overlap, which leaves plenty of room for confusion. Here's a quick review.

Crowdsourcing is "the act of outsourcing tasks, traditionally performed by an employee or contractor, to a large group of people or community (a crowd), through an open call," according to Wikipedia. Crowdsourcing should not be confused with open innovation, but rather thought of as one tool that can be used to bring external input into your organizations as your pursue open innovation.

Many companies mistakenly seem to view crowdsourcing more as a marketing tool than as an open innovation tool. Crowdsourcing gives companies great opportunities to interact with customers and users. Using crowdsourcing, companies can collect ideas, let the crowd vote on them, and set up prediction markets that can bring outside input to any number of decisions. Crowdsourcing has enormous appeal to savvy marketing executives who understand the power of social media. While the marketing benefits alone are often enough to justify the launch of crowdsourcing initiatives, the real innovation opportunities possible are quite valuable as well.

User-driven innovation, on the other hand, is a technique in which companies observe their users in order to gain new insights that can be used in the innovation process. Some confuse this with open innovation because they believe everything that is done with users or customers constitutes open innovation. However, it does not become open innovation until you really start opening up throughout the whole innovation process. It is not open innovation if you just observe others in the early stages of the innovation process—although this can bring much value—and then do everything internally in the same way as you have always done it.

Co-creation is probably the term that is closest to open innovation although it seems to focus more on co-creation with customers rather than with other companies in an ecosystem [discussed below]. Co-creation is viewed as a process in which companies and active customers share, combine, and renew each other's resources and capabilities to create value through new forms of interaction, service, and learning mechanisms. It differs from the traditional active firm–passive consumer market construct of the past. I like how C. K. Prahalad and Venkat Ramaswamy argue that "value will be increasingly co-created by the firm and the customer…rather than being created entirely inside the firm."[2] Co-creation in their view not only describes a trend of jointly creating products, it also describes a movement away from customers buying products and services as transactions, to those purchases being made as part of an experience. The authors hold that consumers seek freedom of choice to interact with the firm through a range of experiences. Customers want to define choices in a manner that reflects their view of value, and they want to interact and transact in their preferred language and style.

Another term I'll use often in this book is ecosystems. With open innovation comes the need to create value networks that include all the potential categories of external sources that can support your innovation effort. Your ecosystem may include customers, suppliers, academic institutions, innovation marketplaces or intermediaries, private labs, government-supported labs and research institutes, government-supported innovation initiatives, and even competitors. At first, this list may seem overwhelming for a small company, but understand that your ecosystem will start small and grow gradually overtime. It may eventually become significantly more complex but that complexity also brings with it a wealth of opportunities. As I will describe in detail later, the ability to build and cultivate good relationships within your ecosystem is a key component of open innovation success.

Why Open Innovation?

Once they get an idea of what open innovation is, many people conclude that it is the Holy Grail and jump aboard the bandwagon without asking this all-important question: Why is open innovation relevant to your company, its present situation, and its mission and vision? If you haven't answered this question thoroughly, you need to bring your feet back on the ground and remember that innovation of any type is just a tool, not a goal. The goal is to grow your company and make a profit.

An answer to the "why" question should show an understanding of how open innovation can be an important part of the general innovation strategy, which in turn needs to be strongly aligned with the overall corporate strategy. But many companies, especially small ones, don't even have an overall innovation strategy, much less a specific open innovation strategy that links with it.

Definitions

The benefit of having an innovation strategy is that it sets a direction for your efforts, hopefully a direction that's aligned with your company's strategy. This also allows you to better define your open innovation in alignment with your strategy. Innovation and—even more so—open innovation can be customized to fit a company's needs in so many different ways. Each company needs to find its own approach, the one that matches its objectives, capabilities, and resources. Once the reason for innovating and the approach to innovation are in place, it becomes easier to work out a strategy and implement it.

Here's an insightful and slightly provocative article by Graham Hill, who is a leading international expert in customer-driven innovation. Although the article is a few years old and thus some of the data is outdated, there still are some good lessons on crowdsourcing in this.

How Understanding Customer Jobs Turns Crowdsourcing into Smartsourcing

By Graham Hill, part consultant, part innovator and part entrepreneur helping companies co-create value with their customers.

> Peter Drucker the gurus' guru famously said, "Because the purpose of business is to create a customer, the business enterprise has two—and only two—basic functions: marketing and innovation. Marketing and innovation produce results; all the rest are costs." Marketing is hard enough to get right, but innovation is a whole lot harder still. Depending on the industry, about 80 percent of new products fail on

introduction in the market. And up to 60 percent fail in reintroduction.

To overcome this disastrous failure rate, companies have started to recruit customers to generate ideas for new products, a process Jeff Howe called crowdsourcing in a 2006 article in Wired magazine. There are a number of great examples of successful crowdsourcing programs, however, two examples illustrate what happens when companies start crowdsourcing programs without really thinking them through properly.

The first of these is Dell with its IdeaStorm program. Anyone can come up with a computer-related idea, post it on the IdeaStorm website, vote for the best ideas, comment about them and hopefully, see them implemented. Sounds great. Why not harness ideas from customers? And why not get customers to vote for them to cut program staff costs.

Unfortunately, crowdsourcing has a number of serious problems. The first problem is that customers, even large numbers of them, typically produce average, unremarkable, incremental innovations, rather than the step-change innovations that companies hope for. Although 12,483 ideas have been posted on the website since IdeaStorm started in February 2007, only 366 have been implemented to-date, a miserly 2.9 percent of the total. And most of the implemented ideas provide only incremental improvements to Dell's business. To its credit, Dell says that IdeaStorm is intended as an extension of its relationship with its customers, rather than just as a source of product ideas. Just as well, as IdeaStorm is a failure as a source of winning new innovations.

The second example is Starbucks with its My Starbucks Idea. Similar to IdeaStorm, My Starbucks Idea allows any registered customer to post an idea, vote for the best ideas, comment on them and see them implemented. Or not as the case may be. My Starbucks Idea, despite receiving over 75,653 ideas, has only implemented 315 ideas to-date, an even more miserly 0.4 percent of the total.

You wouldn't think that having ideas to improve a coffee-house chain would be all that difficult to implement. But the low rate of implementation illustrates the second problem with crowdsourcing; customers have no idea of how the business works, what business capabilities it has and thus, no idea whether even the simplest of ideas can realistically be implemented (let alone whether they will turn a profit).

In stark contrast to Starbucks, Toyota implements over 1,000,000 employee ideas every year, 95 percent of them within 10 days of being submitted. But unlike Starbucks' customers, Toyota's employees know exactly where the best innovation opportunities lie, what can realistically be implemented and the profit-impact of doing so. Coming up with innovations like this is part of the Toyota Way. It's what makes Toyota such a unique company.

And there is another big problem too. Customers expend a lot of creative goodwill generating ideas for Dell and Starbucks, only to see the vast majority of them shot down by their peers or ignored by the companies. All that talk by Dell of building a relationship with customers quickly comes to nothing when its customers' hard work creating ideas is neither recognized nor rewarded.

As Dell and Starbucks attempts at crowdsourcing illustrate only too clearly, crowdsourcing's supposed advantage of harnessing customers as sources of innovation is in fact its biggest weakness. The fact is that most customers simply don't have any good ideas and many of the good ideas that do come forward are simply not implementable. This produces a miniscule implementation rate that burns a lot of customer goodwill in the process.

So what should companies who still want to harness customers to generate winning innovations do? How can they turn wasteful crowdsourcing into productive smartsourcing. One company that got it right is Cisco with its I-Prize competition. Starting in late 2007, Cisco asked innovators to come up with ideas that could it turn into the next billion dollar business. Cisco collected over 1,200 ideas from 2,500 innovators in 104 countries across the globe. The ideas were initially filtered to see if they tackled Cisco's pain paints, if they could be delivered using Cisco's capabilities and if Cisco could make money from doing so.

The filtering was done by a full-time, six-man team, drawn from across Cisco's business. The best 40 ideas were then assigned a mentor to help the innovators turn their idea into a workable business plan. The final 10 ideas were then selected, and taken though an interview and further filtering process to find the eventual winner. A single idea—for a smart electricity grid—by a German/Russian team was selected to collect the $250,000 prize.

Cisco's success at smartsourcing shows the importance of focusing ideas on particular pain points or opportunities. Rather than just let innovators come up with ideas, it is much better to focus their creativity on just those opportunities where they can produce a breakthrough. As innovation gurus Tony Ulwick of Strategyn and Prof. Clayton Christensen have shown, in today's customer-centric business environment this means understanding the jobs customers are trying to do and the outcomes they want from doing them. And not only functional "doing" jobs, but also emotional jobs that describe how the customer feels about what they are doing and social jobs that describe how the customer relates to their peers.

Once you understand customer jobs and outcomes, they should be prioritized to find the areas where the importance to customers is highest but satisfaction with current solutions is lowest. This is the innovation sweet spot. Experience suggests that focusing on the innovation sweet spot can produce an 80 percent success rate for new products; a whole lot better than the 80 percent failure rate we commonly see.

Key Chapter Takeaways

- Open innovation is about bridging internal and external resources and acting on those opportunities. This contrasts with closed innovation, where you do not attempt to assimilate input from outside sources into the innovation process, and you avoid having to share intellectual property or profits with any outside source.

- Open innovation should take place throughout your innovation process, not just in the early phases during what we call the front end of innovation. You will miss out on the full potential of open innovation if you more or less deliberately shut down to external resources later in the process.

- With open innovation comes the need to create value networks—or ecosystems—that include all the potential categories of external sources that can support your innovation effort. This may include customers, suppliers, academic institutions, and even competitors.

- Before jumping into open innovation, determine why open innovation is relevant to your company, its present situation, and its mission and vision.

- Each company needs to find its own approach to open innovation, one that matches its objectives, capabilities and resources.

1. H. W. Chesbrough, *Open Innovation: The New Imperative for Creating and Profiting from Technology* (Boston: Harvard Business School Press, 200(3), xxiv.

2. C. K. Prahalad and Venkat Ramaswamy, "Co-Opting Customer Competence," *Harvard Business Review* (February 2000), http://hbswk.hbs.edu/archive/1299.html.

2
OPEN INNOVATION
BENEFITS
FOR SMALL AND LARGE BUSINESSES

Many people—including those who operate small companies—do not link small business with innovation. Yes, they are entrepreneurial, but this is not necessarily viewed as innovation, as these people tend to think innovation is just for big corporations. Yet, several years ago, when Intuit began sponsoring research that would look at the future of small business, one of the first topics they chose to look at was small business innovation. Their report, "Research Brief: Defining Small Business Innovation," states that "the intensity of economic and technological change means few businesses can succeed without continually incorporating new methods. Innovation, once optional for most small businesses, is now mandatory."[1]

The report goes on to note, however, that many small businesses don't call what they're doing innovation. Instead, they use words like "tweak," "adjust," "improve," and "change" to describe how they are continuously improving their product or service in response to market demands.[2] They think of it more in the vein of normal operations than as being innovative. And of course, they are unlikely to have an R&D department or someone who is in charge of innovation.

Yet, as the Intuit report notes, just because small business owners and managers don't see themselves as innovators doesn't mean they are not innovating. In fact, small businesses use the market as their innovation lab as they constantly respond to the needs of their customers. As small organizations, they generally have very close relations with customers, enabling them to keep a close watch on changing customer needs.

Intuit uncovered three main drivers of innovation in small business[3]:

- **Necessity: No small business can stay alive without constantly changing to meet the shifts in their marketplace. Look back at your product or service or how you operated five years ago or ten years ago and compare that to what you are doing today. You will see that a world of change has occurred in many aspects of your business. So while you may not call what you do innovation, it is still innovation.**

- **Opportunity: Small businesses know how to strike while the iron is hot when it comes to taking advantage of a new opportunity. With relatively flat organizations, decision making is fast and opportunities are quickly seized upon.**

- **Ingenuity: Many small businesses are started by people who are frustrated by not being able to find a product or service to meet a need they have. The ingenuity that enables them to fill that need means they have the creativity to support innovation. In addition, while many people can spot a need,**

it takes a lot of drive and passion to devote yourself to coming up with a new solution to meet that need. This drive and passion are key elements required to be successful at the long, hard slog that innovation often entails.

One more important idea from the Intuit report is their conclusion that innovation is less risky for small businesses than large corporations. "Because of their smaller scale, small businesses can experiment with and implement new approaches faster, easier, and cheaper than large corporations. And unhindered by decision-making bureaucracy and remote decision makers, small businesses can move much more quickly on innovation opportunities."[4]

Open Innovation and Large Companies

If small businesses are natural innovators, where does open innovation fit into the picture for them? Although open innovation was discussed going back as far as the 1960s, it was not until the past decade that it really came alive in large corporations around the world. Leading the way was Procter & Gamble (P&G), the world's largest consumer-packaged goods company. In 2000, its new CEO, A.G. Lafley, expressed the radical idea that half of the company's innovation output should include a key external contribution. This led to the adoption of an innovation model P&G calls Connect + Develop via which P&G accesses externally developed intellectual property in its own markets while allowing its internally developed assets and know-how to be used by others. It collaborates with individuals, companies, laboratories, research institutes, financial institutions, suppliers, academics, and R&D networks.

P&G has had tremendous success with its open innovation efforts, success that has encouraged ever-growing numbers of corporations to follow its lead over the past ten years in search of the benefits that accrue from open innovation. Chief among these benefits––which apply no matter how large or small an organization is––are:

- **Faster development and market launch of new products and services, which will build revenues, market share, and profits;**
- **More diversity brought to innovation, which will result in uncovering more opportunities;**
- **Improved success rate of new products and services by making the innovation process stronger; and**
- **Diversified risks and the sharing of both market and technological uncertainties of innovation.**[5]

Obviously, these are important advantages for organizations of any size, especially in today's hyper-competitive global marketplace. But these are only the beginning of the reasons companies both large and small are embracing open innovation.

In a webinar conducted in advance of Co-Dev 2011, an annual conference on co-development and open innovation, Blaine Childress, a research scientist at Sealed Air Corporation, delineated a number of other important reasons Sealed Air, which is a large corporation, started pursuing open innovation several years ago. These reasons apply equally well—or perhaps even more so—to smaller companies:

- **You can't know everything; the world of science is far too big for any organization to employ experts in every field.**
- **Science and technology advance too rapidly for any one company to cover all advances in any single field.**
- **Innovation is essential for market survival.**
- **Talent is a global resource.**

- **Organizations must fill skill gaps by finding external partners (enlarging networks).**

Let's look more closely at these reasons. First, P&G learned that for every one its 7,500 scientists and researchers, there were 200 others outside their organization who were just as good or better, meaning there are perhaps 1.5 million talented people outside of P&G who could be of value to the company's innovation effort. How many experts do you suppose are out there in the world who have knowledge and ideas that might benefit your small company?

To Childress's idea that science and technology are advancing too rapidly for a company to keep apace, I would add that business and consumer trends are also evolving at a far more rapid pace than ever before. This makes it difficult for even large companies to stay on top of things, never mind the challenge that this presents for smaller companies without access to expensive research that big companies use to track shifts in the marketplace.

Over the past few decades, it has become a given that innovation is necessary for survival. And by this I mean real innovation, not just slapping a new name and packaging on something old. The creation of new value is what separates the companies that thrive from the ones that languish in the doldrums or even fail altogether.

The bankruptcy of Borders Books is an excellent example of what happens when you fail to innovate in a marketplace that is undergoing major changes. Over the past decade, this bookseller lagged far behind its competitors in keeping pace with the changes that are transforming the industry. First, it outsourced its online presence to Amazon.com for a number of years instead of developing a strategy that would enable its own brand to thrive online. With books being the most popular online purchase, not having a viable online brand presence hurt Borders dearly. To make matters worse, the company was extremely slow to embrace the e-book phenomenon. Everyone has heard of Amazon's Kindle and Barnes & Noble's Nook, but how many people know the name of Borders' e-reader? (It's the Kobo.)

As Borders learned the hard way, innovation is imperative in today's marketplace. This is true whether you're in the business-to-business market or serving the needs of consumers. So the question is not whether you have to innovate; it's how you are going to make that innovation happen. And for increasing numbers of companies of all sizes, the answer to that question is through open innovation.

As Blain Childress points out, few companies can afford to hire all the needed expertise internally. This, of course, is especially true for small companies. Yet as he also notes, companies of all sizes now have access to talent from around the world, not just what's in their own neighborhood. And small companies can take advantage of this resource through innovation intermediaries, as described in Chapter 9.

Because of the high level of risk-taking involved with young ventures, leaders of entrepreneurial enterprises often have healthy or even outsized egos; it takes a certain amount of hubris to believe you can defeat the high odds against the success of a new venture. This can lead you to ignore the reality that A. G. Lafley at P&G knew and what Blain Childress talked about: that you and your people may, in fact, *not* have the best ideas and solutions. In reality, there is a strong possibility that the best people and the best ideas are to be found outside your organization. Failure to recognize this can hold a company back from realizing its full potential.

And the problem can only become worse as an entrepreneurial company moves beyond the early phases, when all activities are geared toward executing on a single, great product, idea, or technology. As the company grows, focus tends to shift toward control rather than keeping the visionary thinking and bold approaches that built the company. These must be re-ignited. Open innovation can be the vehicle for accomplishing this objective.

Of course, open innovation offers challenges as well, especially for people who are accustomed to working in a closed innovation environment. These three fundamental questions must be answered by businesses of all sizes before embarking on a journey toward open innovation:

- What will open innovation do to your business model? In an open innovation world, you may end up working with anyone—even competitors. How will this impact your business model and alter your competitive landscape?

- How will your organization change to accommodate open innovation? What kind of collaborations do you want to engage in? What common vision and mission will you share with partners? Systems, processes, values, and culture across the company will need to be transformed. People who have spent their careers being internally focused must now focus externally as well. Change is resisted in many organizations, no matter their size.

- Do you and your organization really understand open innovation and what it takes? Silos can exist even in small businesses, meaning that even a small company may have problems with innovating internally, let alone doing so with outside partners. Leaders need to understand the impact of this movement—its opportunities and threats—and learn to adopt a style that optimizes trust, motivation, and performance. And they need to help their entire team understand the whys and wherefores of open innovation. This may be simpler in a small organization since it probably won't entail educating hundreds of individuals.

Interview #1: LG Experiences With Small Companies

Chris Ryu, Open Innovation Manager at LG shares insights and experiences on how LG innovate with small companies. You can find more information on LG´s open innovation efforts on their Collaborate and Innovate portal.

How well informed do you find small companies to be about open innovation?

Ryu: From my experiences, they're not informed with the term, "open innovation", but I find that they understand and appreciate big companies' open innovation initiatives. And for them, developing internally is not an option, so they must find right partners to work with to launch their products or solutions.

What have been your key learnings in working with small companies on open innovation projects?

Ryu: A key learning is that unless a large majority of employees at your company understand and practice true open innovation, you need to have dedicated teams to drive open innovation activities, and coordinate collaborations between you and small companies.

Small companies don't have the resources to find the right contact in big companies and as a result they often end up continuing to negotiate with people who have a closed innovation mindset.

What advice would you give small companies that want to innovate with big companies?

Ryu: The most important thing is to find the right contact. You have to find the right people on both the working- and management-level. People at big companies tend to get busy with other stuff, so you need to make sure to keep contacting and updating them about your current status and issues.

It takes a lot of resources, but when you try to solve issues, it is always better to meet in person. Often, exchanging emails send the wrong messages to people. This is especially true for non-native English speakers.

It might also be a good idea to use local agency that understands culture and language if you're introducing your technologies to overseas companies. Often, people from small companies who deal with big companies are not good at managing relationship between companies. These coordinators need to excel at building relationships, keeping constant communications, negotiating, culture, language, etc.

How can big companies improve their ways of innovating together with small companies?

Ryu: The only way to improve is to have a strong commitment from the top managements and a strong open innovation team that drives the corporate culture.

What are the benefits of innovating with small companies?

Ryu: Big companies cannot develop everything by themselves. Also big companies can't specialize in many areas.

Are you capable of detecting early signs of danger or promise when you engage with small companies? If so, what are they?

Ryu: I tend to be careful when they are very stubborn and thus hard to work with. I also pay attention when they have very poor presentation skills and when I cannot see the clear value in their technologies or services.

What are the key elements needed to create win-win situations?

Ryu: You need very clear and frequent communication. This is extremely important and a face-to-face meeting always helps. It is a good idea to have a mediator who can coordinate collaboration between big companies' labs and small companies. We need to remember that negotiation is not about "we win and you lose," but about coming up with creative thinking to solve conflicts.

Small companies often have limited legal resources. What can they do to get better deals and protect their intellectual property?

Ryu: They should sign an NDA if confidential information is to be discussed and they should always thoroughly read the agreements even if they are under pressure from big companies to review faster.

Interview #2: Arcimoto and Open Innovation: Don't Reinvent the Wheel

The mission of Arcimoto is to develop ultra-efficient mobility solutions—vehicles, electronics, and software—to catalyze the shift to a sustainable transportation system. Arcimoto's first product, Pulse LT, is an ultra-efficient electric commuter vehicle that strikes a balance between utility, economy, and pure electric fun. Arcimoto is working on the cutting edge in many ways and I wondered what open innovation means to a company like this, so I interviewed serial entrepreneur Mark Frohnmayer, who is Arcimoto's founder and president.

Mark started out by stating that, "as a vehicle developer, Arcimoto works with a large collection of suppliers, large and small, to develop our products. Since our primary role is design and integration, an open, collaborative approach is a fundamental requirement at least to a degree. As we plan the rollout of our platform strategy, we're also looking to 'available source' software models as well to continue to push the boundaries of openness."

What does open innovation mean to Arcimoto?

Frohnmayer: I had to look up the term on Wikipedia for the common definition. To me, open innovation means focused inter-organizational collaboration and to some degree the willingness to engage the masses (crowdsourcing, etc.) to participate in the innovation process.

Can you share how Arcimoto benefits from innovating with others?

Frohnmayer: We don't have to reinvent the wheel. Literally. There are a lot of good wheels out there, and brake lines, and motors and batteries and controllers, and so on. Partnering with other organizations lets us leverage the advances others make into improvements in our own product offerings. We can focus on where we actually add value.

A smaller company like Arcimoto is often faster and more agile compared to big companies. Which benefits or opportunities can Arcimoto bring to bigger companies in a potential partnership?

Frohnmayer: Depends on the company. Materials suppliers can test cutting-edge materials on small production runs with much greater time efficiency than an automotive line that has to scale quickly to hundreds of thousands of units. From a purchasing angle, Arcimoto's products are tuned to the needs of most commuting drivers, so large companies can easily integrate Arcimoto products into vehicle fleets and meet ever-increasing efficiency goals.

What concerns should small companies have about open innovation?

Frohnmayer: Being clear about where they are really adding value, and making sure such value is properly rewarded and protected.

What kind of people work with open innovation at Arcimoto? Do they have specific traits, skills, or mindsets?

Frohnmayer: The engineering team is the primary group within Arcimoto currently innovating with outside partners. Communication abilities are key, as well as a willingness to look outside of the organization for a solution.

Small companies often have limited legal resources. What actions has Arcimoto taken to get better deals and protect your intellectual property?

Frohnmayer: Like you said, small companies often have limited legal resources. For us that has meant having to prioritize which elements we choose to protect as intellectual property. We've filed for provisional patent protection for several elements of the vehicle architecture, and will be patenting the industrial design of the product as well. Although I see patents as useful holdings, our main focus is building the brand of the company and the products it represents.

Key Chapter Takeaways

- Just because small companies don't see themselves as innovators doesn't mean they are not innovators. In fact, they use the market as their innovation lab as they constantly respond to the needs of their customers.

- According to Intuit, necessity, opportunity, and ingenuity drive innovation in small businesses.

- Open innovation can speed the development and market launch of new products and services, bring more diversity to innovation, improve the success rate of new products and services, and diversify risks and share both market and technological uncertainties of innovation.

- Open innovation addresses the issues posed by rapidly advancing science and technology and opens up your company to talent from around the world, enabling you to fill gaps in expertise.

- The pursuit of newness through real innovation is what separates the companies that thrive from the ones that languish in the doldrums or even fail altogether.

- There is a strong possibility that the best people and the best ideas are to be found outside your organization. Failure to recognize this can hold a company back from realizing its full potential.

- Before embarking on a journey toward open innovation, you must explore what open innovation will do to your business model, understand how your organization will change to accommodate open innovation, and determine whether you and your organization really understand open innovation and what it takes.

1 "Research Brief: Defining Small Business Innovation," *Intuit Future of Small Business Report* (March 2009): 1, http://http download.intuit.com/http.intuit/CMO/intuit/futureofsmallbusiness/

intuit_fosb_report_march_2009.pdf.

2 Ibid., 2.

3 Ibid., 3.

4 Ibid., 5.

5 M. M. Keupp and O. Gassmann, "Determinants and Archetype Users of Open Innovation," *R&D Manage* 39 (2009): 331–41.

3
THE INNOVATION ECOSYSTEM
WHEN BIG COMPANIES MEET SMALL COMPANIES IN OPEN INNOVATION EFFORTS

Open innovation has been a hot topic among larger companies for a while now. This is true even in more conservative industries that have higher barriers to adopting open innovation, which include industries characterized by high capital investments, long product development cycles, and stronger focus on intellectual property rights (IPR) issues. Imagine a continuum with two opposite ends. At one end, we have fast-moving consumer goods companies, many of whom have been early adapters of open innovation. At the other end, we have pharmaceutical or semiconductors companies, among others, which must deal with the higher barriers mentioned above. Yet even innovation directors in companies in such industries are trying to figure out how they can embrace open innovation. And certainly large corporations in industries without such constraints are exploring open innovation or have already adopted it as a part of their innovation focus.

As more and more big companies embrace open innovation, opportunities are created for small companies to become part of these corporations' open innovation ecosystems, a concept I defined back in Chapter 1. Right now, there is a lot of talk about becoming the preferred partner of choice within innovation ecosystems and thus grabbing innovation leadership positions. Many are still in play, but they are being locked down fast. Corporations need to understand this.

Corporations that are taking the lead in open innovation are hoping to gain an advantage over their competitors as they will get access to a more diverse inflow of opportunities, which can lead to faster and better innovation. Specifically, as they look to bring smaller companies within the orbit of their open innovation programs, corporations understand that small companies bring these advantages to the table:

- **Small companies are often at the leading edge of breakthrough or disruptive innovation. Breakthrough innovation—that is, innovation with potential to be a real game changer—can be exceedingly hard to achieve in a large, bureaucratic organization where people work in silos, have their own turf to protect, and are wedded to the status quo. In contrast, entrepreneurial companies are often founded based on a breakthrough idea that the founders are passionate about and deeply committed to. They may lack the resources to bring this breakthrough to market, however, so they need the resources the larger company can provide.**

- **Small companies can take risks that large companies can't afford to take. Bigger entities have to protect and defend** their established core business operations, so the price of failure for the small, agile start-up is significantly less than that of a large corporation. At this level, people tend to embrace risk, while the larger companies may have cultures that don't support risk taking at all.

- Smaller companies are often closer to the markets they serve than large corporations are to their markets. As a result, small companies can be effective in helping large companies obtain a better grasp of changing needs within a market and better insights into innovations that might meet those needs. Smaller companies may also have developed ties with submarkets that corporations have not been able to reach. This again offers more opportunities for innovation.

- Smaller companies are often more agile than large corporations. The approach and mindset of those operating in small businesses can provide a breath of fresh air to large corporations that are set in their ways, bound by tradition and afraid of change. Innovation requires agility, something many large corporations lack and many small companies have in abundance.

Challenges Abound

While working with small companies has strong benefits, it is not without its challenges for large organizations. Open innovation is a paradigm shift in which companies must become much better at combining internal and external resources in their innovation process and acting on the opportunities this creates.

If you want to bring external partners into your innovation process, these partners expect you to have your own house in order. If you fail to work efficiently with these partners, nothing happens in terms of outcome. Even worse, the word spreads that you are not a good innovation partner and thus you will have a harder time attracting future partners.

Some large companies believe that if they just embrace open innovation, then all their internal innovation issues will be solved. This will not happen. Open innovation is not a holy grail. Open innovation is very much about managing change. If a company can handle the change process related to implementing open innovation, then they will have learned valuable lessons that can be used in change management situations. In the current and future business climate, I think everyone should appreciate working in an organization that is agile and prepared for changes.

Often, the biggest enemy of innovation is the company itself as it begins to focus more on its own needs rather than the needs of the market. When you begin to innovate with partners, you will see that these partners either focus on their own needs—and then innovation will definitely fail—or you will see that they come together and funnel their resources toward a market need. If the latter happens, then you have a great chance to succeed with innovation. Pressure from external partners can shift awareness from internal needs to market needs and this move can be helpful beyond the innovation process.

Open innovation can bring along new organizational structures. As open innovation becomes the way to innovate, the functional/divisional or matrix organizational structures as we know them today will change—or perhaps even break down. This can be extremely scary for the lovers of the status quo within a corporation. Resistance may be steep within such organizations unless leadership plays a strong role in leading such change.

The increased number of people involved in open innovation provides new ways for people to connect and therefore new ways to be creative. This can also increase the level of complexity.

Culture and Mindset

Anyone who has worked for both a large corporation and a small, entrepreneurial company can talk endlessly about the differences in the two cultures and mindsets. In response to a blog post I'd written about why small companies should embrace open innovation, Russ Conser from Shell's GameChanger program had some interesting comments about the cultural differences. Here's what he said:

> Stefan, as is often the case, you're on the right track—SMEs need big companies as much as big companies need SMEs. The trouble is that it's a classic example of one being from Venus and the other from Mars. Precisely because they're different and they need each other, they often come into conflict early in a relationship. Us big-company guys still have plenty to learn about how to work with SMEs without killing the very things that make them valuable (e.g. creativity, flexibility). Meanwhile, SMEs also need to learn to have realistic expectations of what they have and what they still need if they want to make their ideas real—much of which can often be found in the hands of the corporate types (e.g. financial resources, complementary skills, or practical context).
>
> My experience is that most often there's more than enough value at the intersection for both to be happy with the gains of a relationship, as long as they don't kill each other in the process of creating it.

Here's another comment worth noting on culture from an anonymous poster in response to one of my blog posts: "Smaller firms tend to be younger and more entrepreneurial. The 'can do' spirit of innovation is often alive and well and a key part of their DNA (think

Silicon Valley start-ups vs. Fortune 500s. As firms grow, this creative energy gets slowly squeezed out as the corporate focus turns to quarterly performance (vs. a longer-term view), stable operations (vs. disruptive innovations), and productivity improvements (vs. investing in growth)."

Stark Differences

The differences that can cause problems when big and small companies come together for open innovation can be stark. Let's look at a few that impact the way the two types of organizations approach open innovation:

- **Speed of decision-making: Large corporations, with their abundance of silos and bureaucratic levels, often require considerable time to make decisions. Analysis paralysis is not uncommon, with decisions that seem simple to an outsider taking ages to make. In contrast, in smaller organizations, decision making can be fairly rapid.**

Thus, when these two types of organizations come together in open innovation, the smaller company may find the speed of progress frustratingly slow. At the same time, the people from the large corporation may be troubled by the constant pleas of the smaller partner to move faster. Both sides may be left feeling that the other side just doesn't understand them.

Intuit, a California-based maker of financial software, is one corporation that has worked hard to overcome this problem. They understand that the reply time they can offer potential partners in their ecosystems is critical. As a result of this understanding, they try to provide a clear go/no-go within just forty-eight hours when they stage their Entrepreneur Days. This can take weeks or even months for many other companies.

- **Attitude toward risk:** How large and small companies feel about risk-taking can vary considerably. Particularly where the smaller company is a start-up or still in a fast-growth stage, the organization at all levels may wholly embrace risk because, at this point, the whole business is a risk. However, in a large corporation that has been around for decades, people may be far more vested in keeping things as they've always been than they are in trying something new and potentially risky. Here again, this difference can lead to frustration on both sides when two such organizations engage in an open innovation partnership.

Here's another aspect of this difference as described by Bengt Järrehult, director of Innovation and Knowledge Management at SCA Hygiene Products and SCA Packaging, in a comment to my blog post on this topic: "The biggest difference as I see it is the balance between the defensive and offensive behavior. The bigger (and often the more mature) the company is, the more the company has to lose, and as 'losses loom larger than gains' (Tversky & Kahneman) the behavior of the larger, mature company gets more defensive and hence the relative focus on more potentially breakthrough innovations versus continuous improvements decreases."

Obviously, if a large company and its smaller open innovation partner differ in terms of what type of innovation they should seek—breakthrough versus incremental—this can cause their partnership to fail. This is why clarity of purpose from the get-go is so important.

Edward Thompson, an advisor in the telecommunications industry, also weighed in on the topic of risk:

"Big companies certainly can innovate. They have the required resources and deep talent pools. However, big companies are often very risk adverse, so it's difficult to get a particular business unit or division to adopt an innovation that is not exactly in their market or technology sweet spot. True innovation often requires a company to embrace a totally new market space. Big companies are very reluctant to take any sort of risk associated with entering a new field."

- **Allocation of resources: In a small company, every penny counts. Resources, which can be scarce, are allocated based almost solely on whether they will boost the bottom line. This bottom line focus may not be so distinct in a larger corporation. With more abundant resources—at least in comparison to smaller companies—people in corporations may be relatively free spenders, although this is certainly not always the case and hasn't been in recent years as the recession has taken its toll. However, the small company may expect its larger partner to foot every bill and may not understand that even big companies have their limits. The result of such a relationship can be similar to problems that arise when two people with very different attitudes toward money and spending get married.**

- **Who understands the business model and who manages it: Here's an astute comment Michael Lachapelle, a Canadian expert on business design and business models, made in response to a blog I wrote about the differences between large and small companies when it comes to innovation:**

One of the considerations in driving innovation is who understands the business model of the company. In a small company it is much more likely that everyone in the company understands how the company works and how the individual parts will combine in the business model to create and deliver value to the customer/client. Larger corporations tend to be much more fractured, and thus the staff is less likely to understand the whole. In this context innovation affecting the whole company can be a hard, long-term task, as one has to build a common understanding and mobilize around very different views of the company. Innovation is more likely to occur at business or product line level, then at a whole of company level.

A second consideration is who manages the business model. In big corporations, people feel responsible for only their portion, or sphere, of control. This control is manifested in being able to influence decisions or budgets, or being able to define meaning within the context of the business. There are only a few people who "control" the whole organization, so innovation of the company has to be managed and driven by the senior executives. The more "distributed" control over the key areas of decisions, budgets, and meaning, the more difficult it is to drive innovation.

Both of Michael's points can lead to frustration for smaller companies when they try to engage with larger corporations in open innovation. Small companies may sense a lack of passion among corporate employees who don't understand the whole business model, and the layering of control in corporations will often lead to slow decision making, as already mentioned.

- **Processes or lack thereof: Many small companies don't yet have defined processes in place to drive innovation forward. This is one of the areas where partnering with a larger company can really benefit them. Ironically, however, as someone pointed out in a comment on my blog, while "in theory, this dimension should go to the big guys who tend to take the time and effort to embed systematic processes," this isn't always the advantage that it might first at seem. He continued, "I'm a big believer that innovation can benefit** from process, tools, and governance and is not just a matter of divine inspiration—although that helps too! However, in practice, not many big firms have GOOD radical/blue ocean innovation processes. They tend to be good at incremental innovation (smaller-scale product improvements or extensions) because (1) this is their bread and butter and so they have dedicated resources focused on this, and (2) radical innovation is high risk and highly disruptive to a large organization from a resource, capital, and management focus perspective."

- **Following rules versus breaking rules:** Mark Palmer, a communication advisor, added that "big companies preoccupy themselves with competitors, the market, and the rules. Small companies are more inclined to make up new rules." This relates to some of the benefits of working with small companies as mentioned earlier. Of course, rules should be followed, but sometimes they do need to be bend—or perhaps even be broken—to make real progress.

Those large firms that are still strong innovators (Google, Apple, P&G, 3M, etc.) tend to be the exceptions that have continued to foster a great culture of innovation while also embedding strong processes that help nurture breakthrough ideas in the challenging confines of large, bureaucratic structures.

At an open innovation conference I attended, Cisco said it is trying to move from a culture of competition to a culture of shared goals, a move largely driven by a desire to innovate with external partners. Open innovation has the potential to change much-outdated corporate thinking beyond the "not-invented-here" culture.

As you work with external partners, you are exposed to other ways of getting things done. You bring diverse thinking into the organization. This can make you consider whether your current practices are good enough, or whether you have to adjust these or perhaps even develop new next practices for your organization. An example: you get new perspectives on collaboration. Perhaps this can inspire to better interaction and collaboration between business units.

That's said, large companies have always used their size and power to get things their way. This is no different with open innovation. So I am not surprised when I listen to people from smaller companies complain about the behaviors of large companies when they start working together.

Such behaviors were confirmed by several large companies at an Open Innovation Summit I attended some time ago. There, executives from large corporations openly shared stories on how they had used their size and corporate power to get deals that favored themselves, and they even admitted that some deals could be so lopsided that they could discourage other smaller companies from working with them. Only half-joking, they also said that their view of win-win situations goes like this: "We win a lot. You win a little." Or, "Win-win means that we get to kick the little guy twice."

At least some larger companies are aware of this situation. I predict that as more large and small companies innovate together, larger companies will realize they can no longer afford to tarnish their reputation by behaving badly in such partnerships. They'll begin to understand how important it is to be perceived as the preferred partner within their industry.

- **Differing definitions of innovation: In response to a blog post I wrote about the differences between large and small companies, Jackie Hutter, CEO and IP strategist at Evgentech, a company with patent pending technology that greatly improves battery charging speed, made the following insightful comment based on her experience:**

One thing I have noticed with my own company is that there can be a real disconnect between the meaning of "innovation" between small and large companies. Some—perhaps most—large companies view innovation as a "super" product development team. These are the folks who look for breakthroughs, but these so-called "breakthroughs" are expected to slot into an existing product or project at the company. In contrast, at a start-up the "innovation" is their whole business, and developed wholly independently of the products and timelines of the large company.

We have already seen this when a Fortune 500 company selected us through an open innovation portal. The company was supposedly looking for innovations like our technology, but they effectively wanted it to fit into their existing infrastructure without much modification. They didn't see that our technology could change the game for them because they were playing a different game than us. But even if they had seen that we were different, they wouldn't have cared because they liked the game they were playing.

Fortunately, we were able to see that this company was not going to be a good partner, and we didn't burn much time on them and moved on quickly. (They seemed a bit surprised when we effectively told them, "we're not that into you.") We learned much from this experience, and have modified our potential partner intake to specify "strong innovation mindset" because we recognize that unless a company is already wired to understand the opportunities that our technology will provide their company, they have a low likelihood of being successful in getting to market with our disruptive technology. And, if they aren't successful, we won't get paid.

Inspiration from an Open Innovation Leader

You can also find some inspiration from an October 27, 2010, blog post from P&G that announces new goals for their open innovation program. It shows how committed the open innovation pioneer has become toward the idea of becoming the preferred partner of choice. I hope this P&G approach can inspire other companies to take the preferred partner of choice idea seriously.

P&G: New Goals for C+D to Accelerate Our Innovation

P&G has set two new goals for our open innovation program with the aim of taking Connect + Develop to the next level of strategic value creation. In short, we are looking to drive greater value from our external partnerships. This will drive a sharper culture change for us. And, we think, create even greater opportunity for our partners. The goals call for

- **C+D to triple its contribution to P&G's innovation development by delivering $3 billion toward the Company's annual sales growth.**

- **P&G to become the Partner of Choice for innovation collaboration by consistently delivering win-win relationships.**

For potential partners, that means we will aggressively seek game-changing ideas to build the business that will deliver greater value and scale opportunity to touch more consumers.

Internally, that means we are going to focus even more sharply on ensuring that we include C+D in our largest initiatives, bringing the best minds to work against the biggest projects to deliver the biggest game-changing wins for consumers.

The second part of the goal is good news, too, as it clearly states that only win-win partnerships are on our radar. We've been committed to this from the very start of C+D in 2001. But now it's an established goal that we want to be THE partner that innovators come to. And they'll only continue coming to us if they are confident that they'll get a fair shake and good value.

C+D already has delivered strategic value all across the Company, and a series of game-changing consumer innovations. That was in the first 10 years of the program. Let's see what we can do now.....[1]

Key Chapter Takeaways

- Small companies bring these advantages to corporations that are striving to become preferred partners of choice within innovation ecosystems: (1) Small companies often are at the leading edge of breakthrough or disruptive innovation. (2) Small companies can take risks that large companies can't afford to take because the bigger entities have to protect and defend their established core business operations. (3) Smaller companies are often closer to the markets they serve than large corporations are to their markets. (4) The approach and mindset of those operating in small businesses can provide a breath of fresh air to large corporations.

- Open innovation is a paradigm shift in which companies must become much better at combining internal and external resources in their innovation process and act on the opportunities this creates.

- If you want to bring in external partners to your innovation process, these partners expect you to have your own house in order. Adapting open innovation will not eliminate your internal innovation problems; you need to solve those first.

- When you begin to innovate with partners, you will see that these partners either focus on their own needs—and then innovation will definitely fail—or you will see that they come together and funnel their resources toward a market need. If the latter happens, then you have a great chance to succeed with innovation.

- When small and large companies intersect for open innovation, they need to overcome differences that may include speed of decision making, attitudes toward risk, willingness to develop new rules instead of following the old ones, allocation of resources, definitions of innovation, and varying processes or a lack of processes. They also need to be clear about each other's business model and who is in charge.

- As you work with external partners, you are exposed to other ways of getting things done. You bring diverse thinking into the organization. This can make you consider whether your current practices are good enough, or whether you have to adjust these or perhaps even develop new next practices for your organization.

- Large companies that hope to succeed in becoming partners of choice must do their best not to use their size to get their way all the time in open innovation partnerships.

1 "P&G: New Goals for C+D to Accelerate Our Innovation," P&G Views *(blog)* Oct. 27, 2010), http://www.pg.com/en_US/news_views/blog_posts/2010/oct/accelerate_our_innovation.shtml.

4
GETTING YOUR ORGANIZATION
READY FOR OPEN INNOVATION

If what you've read about open innovation so far makes you think your company could benefit from it, you're probably wondering what you need to do to prepare for making open innovation part of your growth strategy. As I mentioned back in Chapter 1, every company, regardless of size, that considers embarking on an open innovation journey should first answer this all-important question: Why is open innovation relevant to your company, its present situation, and its mission and vision? If you haven't answered this question thoroughly, you need to remember that open innovation is just a tool, not a goal. The goal is to grow your company and make a profit.

An answer to the "why" question should show an understanding of how open innovation can be an important part of the general innovation strategy, which in turn needs to be highly aligned with your overall corporate strategy. But many companies, both small and large, don't even have an overall innovation strategy, much less a specific open innovation strategy that links with it.

Also, I recognize that strategic planning itself is not a strong point for many small companies. They're often too busy keeping their heads above water to engage in what should be a strong, annual process. So if your strategic plan has been gathering dust on the shelf for several years, before trying to answer questions about your innovation strategy and your open innovation strategy, you'll need to update your plan to make sure your strategy still fits with shifts that have occurred in your marketplace as a result of competitive or economic changes or technological advances.

A Holistic View

As mentioned earlier, the benefit of having an innovation strategy is that it sets a direction for your efforts. This also allows you to better define open innovation in terms of your company. As I also mentioned earlier, there is no single method for open innovation. It can work in many different ways and involve a number of different types of partners. The possibilities are almost endless. Finding an approach that meets your company's individual needs, resources, and market situation is essential. Once you've figured out why you need open innovation and how you're going to approach it for your particular company, it becomes easier to work out a strategy and implement it. The key thing here is the ability to view innovation in more holistic terms. Innovation should be about more than just core products and services, and it should involve as many business functions as possible since innovation that supports the growth of your company can come from almost any part of your company. Open innovation gives you a chance to break down operational silos, which exist even in small companies.

Other Necessary Elements

Your next steps will include putting in place these five key elements that are necessary for making any open innovation strategy a success:

Element #1: Stakeholder Analysis

Implementing open innovation is a paradigm shift that requires people to change their mindsets and obtain new skills. So it's critical to start with an overview of your internal and external stakeholders. Who will be affected by the open innovation intention? Analyze the pros and cons of your open innovation initiative for these people. What issues bother these people? How can you create a value proposition that will make the stakeholders support the initiative?

One approach is to create a stakeholder map that identifies all the various groups that might be impacted by your open innovation initiative, then develop specific value propositions for each group. Don't forget to focus on informal influencers, that is, people with a disproportionate level of influence. If you are not familiar with network analysis where this is a common concept, then you should imagine that any organization has two kinds of organizational hierarchies. One is the formal hierarchy where you know who the managers are by their titles and job descriptions. The other is the informal hierarchy built on the trust and respect among the employees. Some colleagues are more trusted and respected than others and as such they are key players in any kind of change management initiative, including the implementation of open innovation. Find these people and win them over to your cause, and it will be easier to build the innovation DNA. You can read more about stakeholder analysis in Chapter 6.

Element #2: Communication Strategy

Strong communication programs are important at any time within an organization, but never more so than when open innovation is your goal. Without a good communication strategy, your odds for success do not look good. To move forward, people need to know where to go and how to get there. Here are three reasons why communication is a critical element if you want to succeed with open innovation.

- "We are in the matchmaking business." This quote came from Chris Thoen, former managing director of the Global Open Innovation office at P&G, during his presentation at a CoDev conference. Chris also stated that one of the key objectives of open innovation is to become the preferred partner of choice. As with any kind of matchmaking, we strive to find the best possible partner and in order to do so we must be able to articulate our propositions in an attractive manner. This is very much about communication.

- Find and be found. Let the ecosystem in your region—as well as globally—know what you are doing. Tell about your open innovation initiatives, share your learnings, and ask for input. Messages with substance move very fast within such ecosystems. This can help a company to be perceived as a preferred partner of choice or at least as a company with a potential for this to happen. You need to find the right partners, but it would be great if they also came to you, right? This is very much about communication.

- Open innovation needs to become top of mind within organizations; not just within innovation teams. By now, many innovation teams understand the value of open innovation and those that do not will soon learn the hard way. It is a tougher challenge making the rest of the company fully understand and buy into the value of open innova-

tion. Nevertheless, this has to happen in order for them to change mindsets and behaviors and thus be able to fully support open innovation and benefit from this. This is very much about communication.

As they embrace open innovation, companies are increasingly realizing that they need to change from just branding and promoting the innovation outcome to doing the same for their corporation innovation capabilities. Doing so helps them build their image as an attractive partner and thus supports the goal of building a strong open innovation ecosystem. Obviously, communications comes into play here, too, both with internal and external audiences.

Knowing the strategic goals motivates people and builds a collective sense of purpose. So put a communications plan in place before you even start. Make sure you take every opportunity to turn good news into a story that can work internally as well as externally. The latter is especially useful if recruitment is a serious issue.

Obviously, communicating within a small company is easier than in a large corporation. But that doesn't mean that developing a solid communications plan is any less important. In a small company, it might be easier for one or two people to throw a wrench into the works, so it's important that everyone understand why open innovation is important to your company's future and what they need to do to get on board.

You should develop specific communication points of view for your stakeholder groups. It's even better if these points of view are aligned with the value propositions.

In a presentation I heard by Jeff Boehm, a former VP of Invention Machine, the innovation software company, he very smartly outlined a four-step process for keeping the need for focusing on innovation constantly in the forefront of employees' minds through smart communications. Here are his steps with some elaboration from me:

- **Positioning: Make it relevant and show success.** To keep people on track when it comes to pursuing open innovation, they need to understand why it matters to the wellbeing of the organization they're part of. In some ways, this may be easier to do in a small company than in a large corporation, because in a small organization it can be easier for people to see how their work impacts the overall organization's wellbeing. They may feel less like just one cog in a giant wheel than someone who is working in a big global corporation. As a result, they may feel see the direct link between what they do and the success of the business more clearly.

One of the best ways to show why open innovation is relevant is to tout success stories. These won't be available early on but when forward progress is made, make sure everyone knows about it.

- **Promotion: Make it obvious.** As discussed in Chapter 4, having a communications strategy and plan in place to promote open innovation is vital. Constantly reinforcing messages about the purpose of open innovation and the desired outcomes will assure that employees understand just how important it is.
- **Calls to action: Make it easy.** This builds on the above. Keep the messages clear and simple to better allow your people to engage in and thus learn about open innovation.
- **Sustain momentum: Make it stick.** When open innovation is first introduced in a business of any size, it is possible some people may conclude that it is another "flavor of the month" initiative sent down from on high. This is particularly apt to occur if an organization has a track record of introducing

new strategies only to abandon them a short while later. This cannot be allowed to happen with open innovation. Actions from the top must continually show that open innovation is here to stay and will be a continuous focus of company efforts. In other words, do everything possible to build open innovation into your company's culture.

Element #3: Common Language

A key objective of your communication strategy is to develop a shared language about open innovation—and innovation in general—within your company. When everyone uses the same language, it is significantly easier to frame the problems and ideas in ways that everyone can understand and relate to.

Large corporations often develop this common language by bringing in outside experts to educate and train people. You may not be able to afford that but there are other ways to get the job done. For example, you can provide people with books and articles about open innovation and then hold discussions to pull out key terms and their definitions that people should become familiar with. Make sure you and your leadership team use this language in company presentations and meetings. Slowly but surely, the language will begin to filter throughout the organization and become part of the company's shared vocabulary.

Some companies develop elevator pitches—very short pitches aimed at getting to the next level of customer contact—for their products and services. Why not do the same for the messages and propositions used toward potential partners as well as the colleagues you need to turn into backers of your open innovation initiatives?

Element #4: Appreciation for Employees

I often say that executives need to create the proper settings and conditions for open innovation since it is so difficult to just implement lots of tools and make open innovation work. One of the key elements of creating the right conditions is to make the employees understand that a stronger focus on external contribution is not a sign of disapproval of the work being done by them. It is merely an attempt to increase the overall innovation productivity by combining internal and external resources. This is equally true for both small and large organizations.

You do not want frustrated employees who do not feel that they are being appreciated.

Having employees who feel appreciated with regards to innovation also makes it easier for an organization to embrace a more holistic approach to innovation in the sense that all business functions should be involved in the innovation process. Innovation is more than just products and technology, and it should not be driven entirely by the R&D function.

Element #5: Networked Innovation Culture

One must-have component of a strong innovation culture is a strong networking culture. To thrive in an innovation environment that becomes increasingly open and externally oriented, people throughout your organization need to be capable of building and sustaining relationships both internally and externally.

Unfortunately, this is a neglected area of attention in companies of all sizes. I remember talking with an executive at a large company after having met with many of the innovation people from his organization. I asked him about his intentions with regard to creating a better networking culture. He did not have any plans. We discussed the benefits of organizational networking analysis that could help identify the people holding important knowledge, as well as the people who were bottlenecks. He saw the relevance. We talked about the importance of being good at relationships in the move toward open innovation. He agreed.

However, when I asked what he would do about the networking culture, I got the same answer. He would do nothing. Why not? "Things like that will take care of themselves," he said.

He could not be more wrong. An innovation culture does not create itself and the same goes for a networking culture. This requires a top-down approach in which you and your leadership team craft a strategy, set the goals, and provide the means and tools for networking initiatives. In addition, you need to find ways to be personally involved in networking to convince employees that you are serious.

It may feel like taking the role of chief networker is adding a huge job to your already crowded to-do list, and it probably is. But open innovation won't happen without a strong networking culture, and serving as a role model is essential.

Why a Networking Culture Is Important

The impetus for creating a networking culture is obvious once you look at the current and future direction of innovation. Let's start by disposing of the myth of the lone genius—the Thomas Edisons and the Alexander Graham Bells of yesteryear—arriving at a breakthrough innovation on his/her own. This model wasn't true then, and even if it were, it simply does not hold true in today's complex business organizations. Technology and the challenges that must be solved have become so complex that many—perhaps even most—companies can no longer rely solely on their own internal innovation geniuses, no matter how brilliant those people may be.

Innovation is increasingly about having groups of people come together to leverage their diverse talents and expertise to solve multifaceted challenges that cross multiple disciplines. To make this happen within your organization—and beyond as you move toward open innovation—requires a networking culture that is designed, supported, and modeled by your company's leaders.

Another key motivation for setting up networking initiatives is based on the simple fact that the knowledge of any company is inside the heads of the employees. Discovering and distributing this knowledge has always been a challenge, and now, more than ever, the ability to leverage a company's collective knowledge and experience is critical to innovation. Furthermore, establishing the ability to bring knowledge and potential new innovation insights in from external sources demands a strong networking culture supported and modeled from the top.

What a Networking Culture Looks Like

So what does a good networking culture looks like? It's such a new concept that there aren't a lot of examples available to illustrate it, but here are some key components of a good networking culture:

- **Top executives have outlined clear strategic reasons why employees need to develop and nurture internal and external relationships. This includes making clear how your company's networking culture links with and supports your innovation strategy.**
- **Among the things to consider when developing your networking culture strategy is what types of networks you hope to build to support your innovation efforts. If your organization is moving toward open innovation, possibilities would include peer-to-peer networks for people working with open**

innovation in different companies, value- and supply-chain networks, feeder networks, and events and forums connecting problem solvers and innovators with your company.

- Leaders show a genuine and highly visible commitment to networking.

- Leaders must walk the walk, not just talk the talk. By making themselves available at networking events and by being visible users of virtual networking tools, they model the desired behavior and motivate others to participate. After all, who doesn't want a chance to exchange ideas with the top brass?

- Leaders should also share examples of their networking experiences whenever possible. Spread the word about your own and others' networking successes. Hearing you talk repeatedly about how networking is helping the organization in its innovation efforts will reinforce the message that this is important.

- Networking initiatives mesh closely with your corporate culture. This is not one-size-fits-all; each company's networking efforts will differ. You can take bits and pieces, concepts and theories, knowledge and experience from others, but you still need to make it work for your own company.

- People are given the time and means to network. Frequent opportunities are provided to help individuals polish their personal networking skills. Not everyone is a natural networker. But almost everyone can become good at it with proper training and encouragement.

- Both virtual and face-to-face networking are encouraged and supported. Web 2.0 tools and facilitated networking events maximize the opportunities people have to initiate and build strong relationships.

Three Types of Networkers

As you determine how to build a networking culture within your organization, it's important to understand how networking actually works. One of the most knowledgeable people on organizational networking—and how it supports innovation—is Rob Cross, an associate professor in the management department of University of Virginia's McIntire School of Commerce. He is also one of the founders of The Network Roundtable. I have learned a lot from my visits and interactions with Rob and I recommend that you take a look at his website at www.robcross.org.

Rob Cross has identified three networking types that you should pay attention to within your organization. They are

- **central connectors;**
- **brokers; and**
- **peripheral people.**

Central connectors are those people with the highest number of direct connections. They can be formal leaders—or political players trying to be leaders—whom everyone seeks out either because they make things happen or because they have made themselves bottlenecks. The latter can become a major issue with regard to innovation, where you often need a dynamic flow. Experts are also a type of central connectors who are sometimes overused, as everyone goes to them with questions. Sometimes experts must be protected.

Brokers are high-leverage employees who connect people across boundaries, such as functions, skills, geography, hierarchy, ethnicity, and gender. These people have the leverage ability to drive change, diffusion, or innovation, and they can also act in key liaison or cross-process roles.

According to Cross, brokers often sit in tipping-point positions and, therefore, diffuse information faster than leaders and central connectors. Even in a small group, brokers are the most effective means to diffuse information. Brokers have ground-level credibility and acknowledged expertise in the eyes of their peers. They are much more likely to be sought out and listened to than a designated expert or leader who might not be influential in the network. Brokers bridge diverse perspectives and understand cultural norms and practices of different groups in ways that those familiar and comfortable within their own group often cannot.

Peripheral people could be new people, experts, sales people, poor performers, or cultural misfits. They sit on the edge of the network, and Cross has learned that typically 30 to 40 percent of peripheral people would like to get better connected but have run into obstacles. They are a resource of untapped expertise but are substantial flight risks. Peripheral people are particularly in need of education about how to network; they will also need more encouragement than others to participate in networking.

Knowing these different types should prompt you to ask questions about what networking types you have within your organization and what type of networking approaches might work best. For example, if your company has a lot of peripheral people, you'll need to devote more time to training in networking skills. Also, as a leader, think about how you can work best with the different types to be able to get the most out of them.

Potential Roadblocks

In working with companies that are trying to build a networking culture, here are some reasons I've identified for why such efforts can fail or not reach the hoped-for degree of success:

- **Lack of time:** Many of us simply do not have the time to network and build relationships. It is necessary to develop a strategy and initiate projects, but you also need to give your people time to invest in initiating and maintaining both internal and external relationships.

- **Lack of skills:** Some people are natural-born networkers; many others are not. But the basics of effective networking can be learned, just like any other business skill. With appropriate instruction and motivation, wallflowers can learn to work a room. By providing your people with this type of training, you will give them a skill that will be invaluable throughout their careers.

- **Lack of focus:** A community or a network will only work if it connects people who share a common experience, passion, interest, affiliation, or goal. Your people need to have ways to find and join groups that are right for them and right for your company. In other words, you and your employees should only network when there is a good reason to do so. Random networking rarely results in anything but wasted time, which devalues networking in people's minds and makes it harder to encourage them to try it again.

When I posed the deliberately provocative question on my blog of whether relationships and networking were overrated, Tim Kastelle, a member of the Technology & Innovation Management Centre in the School of Business at the university of Queensland, commented, "I view most things through a network lens, and the thing that I always remind people is that maintaining links in a network is costly. So just blindly 'networking' to build up connections is likely to do more harm than good. Going down to no connections doesn't work either though. You have to think about the kind and quality of connections that you want."

I fully agree with Tim and recommend using the mantra of "networking with a purpose."

- **Lack of commitment and structure: The networking-will-take-care-of-itself-and-you-do-not-need-to-work-at-it attitude is not the approach to take toward building what is increasingly a core innovation skill. Building a networking culture requires commitment and structure to support it.**

Key Chapter Takeaways

- If your strategic plan is several years old, before trying to answer questions about your innovation strategy and your open innovation, you'll need to update your plan to make sure your strategy still fits with shifts that have occurred in your marketplace as a result of competitive or economic changes or technological advances.

- Finding a definition for open innovation that meets your company's individual needs, resources and market situation is essential. Once you've figured out why you need open innovation and how you're going to define it for your particular company, it becomes easier to work out a strategy and implement it.

- To have your open innovation strategy succeed, you must put in place these five elements: (1) stakeholder analysis, (2) communication strategy, (3) common language, (4) appreciation of employees, and (5) networked innovation culture.

- Keep open innovation in the forefront of employees' minds by (1) making it relevant and showing success, (2) promoting it, and (3) making it easy and (4) making it stick.

- Key components of a good networking culture include (1) clear strategic reasons why employees need to develop and nurture internal and external relationships; (2) an understanding of the types of networks you hope to build to support your innovation efforts; (3) leaders who show a genuine and highly visible commitment to networking, who walk the walk, not just talk the talk, and who share examples of their networking experiences whenever possible; (4) networking initiatives that mesh closely with your corporate culture; and (5) both virtual and face-to-face networking opportunities.

- People must be given the time and means to network and be provided with help to polish their personal networking skills.
- Pay close attention to the three types of networkers in your organization: central connectors, brokers, and peripheral people.
- Avoid these roadblocks to building a networking culture: not enough time or skills and lack of focus, commitment, structure, and communications.

5
GETTING YOUR PEOPLE READY
FOR OPEN INNOVATION

The leaders of successful small companies understand how important it is to have the right people in the right positions. When resources are slim, the ability of everyone to do their job well matters tremendously. One or two weak links can spell the difference between success and failure. So it will come as no surprise when I say that people matter more than ideas when it comes to making innovation of all types happen.

You should take a moment to think about that since many innovation initiatives fail miserably because their leaders don't understand this simple fact. In fact, it is actually more important to have grade-A people than it is to have a slew of grade-A ideas. Why? Because grade-A people can take a grade-B idea—or perhaps even a grade-C idea—and turn it into a successful reality. Grade-B people, on the other hand, will struggle with even truly great ideas.

If we take this to the world of small business, the big question is whether you have enough available grade-A people within your organization who can take great ideas, whether they come from inside or outside the company, and turn them into reality.

When large corporations tackle this question, their answer is simple; with their large body of employees, they can easily switch great people to other projects. But for a small company with its smaller staff, you simply don't have the ability to do that. In this case, it is particularly critical to identify and develop people with the attributes and skills needed to turn an idea into a finished product or service. So before you get all fired up about generating a ton of ideas, first figure out how you're going to match those ideas to people who can make things happen.

Before You Start

As you start this work, here's another key point to remember: The skills needed to lead and manage a project within the existing core business—where innovation is likely to be incremental and resources plentiful—are significantly different from the skills needed to overcome the challenges and obstacles that greet almost any new business project involving breakthrough or radical innovation. And this is especially challenging in small companies where resources may be hard to come by. You need to staff new business projects with people who have a mindset and toolbox that match this different challenge.

You also need different people for the different phases of the innovation process, which presents another challenge for small companies. Just as some entrepreneurs are better at running a company at its very early stage and others are better at helping the business scale once the product is launched, so, too, are there intrapreneurs who are better suited both in terms of mindset and skills to various phases of the innovation process.

For example, the discovery-innovation-acceleration (D-I-A) model of innovation put forward by the Radical Innovation Group identifies three phases of innovation:

Discovery:

- **Basic research: internal and external hunting**
- **Creation, recognition, elaboration, and articulation of opportunities**

Incubation:

- **Application development: technical, market learning, market creation, and strategic domains**
- **Evolving opportunities into business propositions: creating a working hypothesis about what the technology platform could enable in the market, what the market space will ultimately look like, and what the business model will be.**

Acceleration:

- **Early market entry: focus, respond, and invest**
- **Ramping up the fledgling business to a point where it can stand on its own, relative to other business platforms in the ultimate receiving unit**[1]

This model has been used with success at many companies, which have learned that very few people have the skills to move from heading the project in the discovery phase to heading it during the acceleration phase. The challenge this presents for small companies is obvious. With far fewer personnel to choose from, it can be tough to fill all the slots identified in this model. The good thing is that you can identify people with the right mindset (see below) and then start working on their toolbox. Making people more ready for innovation by continuously developing their toolbox is one of the low-hanging fruits and this can be done in small as well as big companies.

The Necessary Skills & Attributes

So as you begin to consider the people on your team, what personal competencies are you looking for when it comes to deciding who can support your open innovation efforts? I was fortunate to see a presentation by Gail Martino, manager of Emerging and Disruptive Innovation at Unilever, in which she identified seven critical personal competencies needed for open innovation success. Here is her list:

1. Intrapreneurial Skills

Sees opportunities and drives an opportunity forward with and within an organization…sometimes before the organization knows they need it.

Key Traits/Behaviors:

- **Aware of and enables new ideas—from anywhere**
- **At ease outside their "comfort zone"**
- **Non-complacent; boundary pusher**
- **Self-driven; risk taker**
- **Not afraid to fail**

2. Talent for Relationship Building Both within and outside the Organization

Key Traits/Behaviors:

- **Is genuine; builds trust**
- **Represents the partner well to the organization**
- **Listener, which is key in building win-win deals**

3. Strategic Influencing

Is able to persuade, inspire, and garner support.

Key Traits/Behaviors:

- **Politically astute; high organizational awareness**
- **Knows when, how, and with whom to gain support**
- **Top-notch communication skills**

4. Ability to be a Quick Study

Self-driven to become up to speed in new areas (critical skill to help showcase and gain

support of others).

Key Traits/Behaviors:

- **Ability to develop expertise, sometimes outside of their area of specialization quickly**
- **Curious**
- **Resourceful**

5. Tolerance for Uncertainty

Able to manage high risk projects and make decisions without perfect information about

outcome.

Key Traits/Behaviors:

- **Manages risks through milestones and keeps moving forward**
- **Doesn't plan for failure—plans to avoid it**

6. Balanced Optimism

Keeps above the fray.

Key Traits/Behaviors:

- **Knowledgeable about risks as well as rewards**
- **Overcomes moods and emotional states to keep moving forward**
- **Non-reactive; evaluates criticism but moves on**

7. Passion

Energy to more forward and overcome roadblocks.

Key Traits/Behaviors:

- **Infectious energy**
- **Sees opportunity, not just issues**

Gail's list expands on the ideas I put forth in *The Open Innovation Revolution* and highlights how important the so-called "soft" skills are when it comes to making open innovation work. As I said at the start of this chapter, because each person involved in a small company's innovation effort is so important due to the small size of the team, you need to carefully evaluate each new hire to see if they have these capabilities. This may require completely redoing your interview process to come up with questions designed to ferret out a candidate's attitude toward risk, for example, or how interested they are in learning about new fields.

At the same time, you need to make sure your company culture supports the capabilities identified above. If every failure is punished, for instance, that will make your company an unwelcome place for the type of risk-takers who are needed to make innovation happen. Here's a quote I often use in my talks that captures what I'm talking about; it's from a blog post written by David Nordfors, founding executive director of the Stanford Center for Innovation and Communication: "When someone tries to innovate within a traditional organization, few will understand what he/she is doing, but everybody will understand who is a troublemaker. After the innovation has been embraced by the organization, few will remember who started it, but everybody will remember who was a troublemaker. This is the dilemma encountered by many intrapreneurs—they risk punishment for success."[2]

Developing the Focus

The challenges of developing open innovation capabilities in a small business are not limited to making sure the right people are in place to move the effort forward. Open innovation team members must be given opportunities to develop the intrapreneurial capabilities discussed above. Also, they need to be constantly reminded why it is important to keep pursuing innovation even when their plates are full with other important duties. Let's look at the latter issue first.

Because employees in small businesses frequently wear many hats, it is all too easy for them to set aside their open innovation hat when pressing matters arise related to the company's core product or service. Put another way, it is not unreasonable for someone to think it's more important to keep the train on the track than it is to devote precious time and scarce resources to trying to improve the train. Countering this natural tendency takes constant vigilance and effort.

Key Chapter Takeaways

- People matter more than ideas when it comes to making innovation of all types happen.

- Your innovation team needs to include people who are good at working in the three-phased discovery-innovation-acceleration (D-I-A) model of innovation put forward by the Radical Innovation Group. Remember that not all people are good at all phases, so you need different people for different phases.

- When looking for people for your team, look for people with these seven skills: (1) intrapreneurial skills, (2) talent for relationship building, (3) strategic influencing, (4) ability to be a quick study (5) balanced optimism, (6) tolerance for uncertainty, and (7) passion.

1 Gina Colarelli O'Connor, "Research Report: Sustaining Breakthrough Innovation," *Research-Technology Management* 52, no 3 (2009): 12-14.

2 David Nordfors, "The Intrapreneur's Dilemma," *The Injo Blog*, Sept. 14, 2008, http://blog.innovationjournalism.org/2008/09/ntrapreneurs-dilemma.html.

6
WHAT TO CONSIDER BEFORE LEAPING
INTO A PARTNERSHIP

You've determined that you want to engage in open innovation and you've developed a strategy, begun to build a networking culture, and are helping your people build the skills they'll need to operate in such a culture. You are now on the verge of moving forward and beginning to engage with other companies. In this chapter I'll talk about how to make such partnerships work. But first, we must look at the one thing that should be uppermost in your mind at this point: how to build the trust that is fundamental to open innovation.

Trust comes at many levels—internally as well as externally. As you move toward open innovation, you should begin to look into two questions:

1. What does it take for you to trust others?

2. How do you convince the people in the organizations you want to draw into your open innovation ecosystem to build trust in you and your company and then start forging strong relationships with them?

The necessity of building trust as a basis for successful open innovation means that it is more relevant to look at the people side of innovation than to concentrate on processes, and it also brings more power to the people who really drive innovation within your organization. Why? Because trust is first and foremost established between people and then perhaps between organizations. Trust is a personal thing, and those who excel at building trust are suddenly in a much better position with regard to making things happen and being valued by their organizations.

It is not surprising that in a Twitter chat I conducted with Chris Thoen, who at the time was running open innovation at P&G, he listed trust and integrity as two of the characteristics P&G looks for in open innovation partners. He also mentioned good chemistry in the working relationship, including the need to have partners who are responsive and who align with you on purpose, values, and principles that guide how a company is run and how people behave as employees. All of this speaks to the importance of the people part of the open innovation equation.

Barriers to Trust

What are the barriers against building trust and strong relationships with stakeholders in your ecosystem?

- **Most organizational structures foster an internal rather than an external perspective.**
- **Most companies view external partners as someone paid to deliver a specific service rather than a source of co-creation and open innovation.**

- **Most companies are more focused on protecting their own knowledge and intellectual property rather than opening up and exploring new opportunities. They play defense rather than offense. This should not come as a surprise, as one of the main objectives for corporate lawyers is to minimize risk, and it is fair to say that opening up to the outside world increases the risk element.**

- **Large companies can be inherently skeptical about the capabilities of small companies. Running into a wall of doubt would not be unusual in these situations. After all, what could a little firm possibly have to offer a global giant? And if the small company's people were really any good, wouldn't they be working for bigger companies to begin with? Yes, I am overstating it here, but you get the point, right?**

- **Forging strong relationships takes time and personal commitment. We are just too busy to make it happen and it does not help that most companies do not provide the necessary time, resources, and encouragement to make this happen.**

What should you do to foster an organizational mindset that supports the building of trust?

The most difficult situation faced by leaders who seek to move an organization toward open innovation is that they are alone. This is a new way of doing things, and it will develop many foes among those who just want things to stay as they have always been. This is true no matter what the size of the company.

This is a very normal reaction; many people feel threatened by something that is new and doesn't seem to match what has led the organization to success in the past. So you do not get much support for this new way of thinking from anyone within your company. They might see that this could be interesting, but once they begin to understand that you have to make significant changes in the way you are dealing with external stakeholders, they begin to raise obstacles rather than see opportunities.

When you have recruited enough people with a proper mindset, then you have laid the foundation for trust, which in turn makes everyone accept that strong relationships are the key to business success in the future. Now you are ready for open innovation.

Unfortunately, it is my experience that few companies have laid this foundation, and this will not happen unless you become successful in recruiting the right people with the right mindset. You can more or less just forget about processes and concepts because, when it comes to open innovation, it is the mindset that matters the most. If you get the mindset right, the implementation of processes will be so much easier to deal with.

Stakeholder Management

Stakeholder management is a critical discipline to master as you try to foster a culture in which trust is a key component and in which resistance to change is minimized so open innovation can thrive. Just as I asked you to consider internal stakeholders when you begin to position your company for open innovation, so you must now consider those same stakeholders and how they might react to the launch of open innovation at your company.

You can get an idea of stakeholder management—for both internal and external stakeholders—by thinking in terms of three steps: identification, profiling, and communication.

1. Identify your stakeholders.

The first step is to figure out who your stakeholders are. Think of internal and external people who can affect your open innovation initiative in both positive and negative ways, and people who might feel threatened or stand to gain from it. Think not only of the obvious people, but also of influencers who are not on a formal organization chart. Prioritize and place important stakeholders on a short list.

As criteria for placing people on the short list, ask yourself two questions: Does this person hold any impact on open innovation in our organization right now? Will this person have a high impact now, soon, or later?

Although working with your stakeholders is important, you will often lack the time to work with all of them, so you need to prioritize them early on. However, you should always be prepared to change the status of the stakeholders and add new stakeholders when you learn of people being affected by your work.

2. Profile your stakeholders.

The next step is to create short profiles of the stakeholders you have placed on your short list. Compile information such as:

- **Orientation: Is the stakeholder internal or external?**
- **General information: What are the name, job function, contact information, and short bio of the stakeholder?**
- **View: Do you see the stakeholder as an advocate, supporter, neutral, critic, or a blocker of open innovation? Why?**
- **Impact: Does the stakeholder have a strong, medium, or weak impact on your open innovation work? Why?**
- **Type of influence: Does the stakeholder hold a formal/direct or an informal/indirect influence on the open innovation initiative? Why?**
- **Key interests: What are the key financial or emotional interests of the stakeholder with regard to your open innovation effort?**
- **The circle of influence: Who influences the stakeholder generally, and who influences the stakeholder's opinion of you?**

To what degree are you connected with the stakeholder and his/her influencers?

3. Communicate with your stakeholders.

The last step is to figure out what you want from your stakeholders and what you can offer them—and then communicate with them.

You might not feel you are ready to do so, but you need to communicate with your stakeholders early and often. This lets them know what you are doing, and you can use their reactions to make changes that can increase the likelihood of success for your open innovation initiative.

People are usually quite open about their views, and the best way to start building successful relationships with your stakeholders is to talk directly with them. You should do your homework before these meetings and interactions. Besides having crafted a profile, you should also know the most compelling messages to use with each stakeholder, and you should be able to deliver quick and concise elevator pitches based on these messages.

It's Homework Time

A stakeholder analysis is not the only homework you need to do when considering open innovation. One of the perilous aspects of open innovation for small companies is the inequity that can come from being in a relationship with a larger company. As the "little guy" it is all too easy to be used and abused by a bigger partner. To avoid entering into a win-lose relationship in which you get the short end of the deal, be sure to take a step I fear too many small companies skip: doing their homework to learn as much as possible about the corporation they're about to join forces with, particularly with regard to that organization's history of dealing with smaller companies. This may be particularly apt to happen if a small company is approached by a larger one instead of the other way around. It can be hugely flattering to have, for example, a Fortune 500 firm say they'd like you to be part of their open innovation ecosystem.

In the rush to say yes, it is all too easy to forget to research your suitor's reputation when it comes to dealing with the smaller fish in their pond. Are they known as the neighborhood bully? It's always a possibility. Even if they are, that doesn't necessarily mean you have to say no. It just means you want to have your eyes wide open from the outset and that you might take more precautions to guard your interests than you would if your research revealed that the corporation that is anxious to work with you has a sterling reputation for how it treats partners of all sizes.

There is a third possible outcome of your research; you may find the larger company has relatively little experience in dealing with firms of your size. This means they may not know much more than you do about how to develop a partnership that will work for both sides. Also, since they're inexperienced at this, they may make some inaccurate assumptions, including thinking that you know how a company of their size operates, when really you do not know this and need to be educated. For instance, they may be used to things taking longer because of their multi-layered bureaucracy whereas you're used to making decisions and moving forward much faster.

Also, if the larger company is new to the open innovation game, they also may tend to assume that they know what's best for you. This may come across as arrogance, when it's really a lack of experience that is causing them to think this way. They just need to learn to listen.

Looking into these issues and others before you leap has other benefits as well. The more you know about a potential open innovation ally from the get-go, the better prepared you will be understand how your company might be of help to the larger company. This means you will be better prepared to showcase your firm's assets. You'll also have knowledge that will be valuable when both sides are trying to set the goals and objectives of the relationship.

Fortunately, an awful lot can be learned from perusing a company's website. How do they talk about open innovation? Or do they even talk about it at all? If they have an open innovation in program in which they're asking people to provide them with ideas, what happens to those ideas? Does their program appear to be well focused, with clear objectives? Or are they just asking for random ideas? (See sidebar for an example of the type of information you can glean from a corporate website that speaks volumes about a company's true commitment to being a trustworthy innovation partner.)

During the research and getting-to-know-you phase, it's good for both sides to look out for hype and exaggerations. Have your "spin" detector on full alert as you listen to a potential partner talk about the wonderful things they can do for you. Ask specific questions about examples of how they've partnered with others and what outcomes were achieved.

It's easier to start a partnership than it is to end it. But don't be so eager that you don't take time to fully consider whether this is the right partner for you. If you're the smaller of the two potential partners, is the other company talking to you as if you're equals or is there an undertone of arrogance? Do you feel you're being talked down to? Are specifics being offered or is everything a little fuzzy? Are goals and objectives clear? Or does it seem that perhaps your prospective partner is getting into open innovation without a lot of forethought, perhaps just doing it because everyone else seems to be doing it? In short, be alert and follow your intuition. If anything feels out of place or amiss now, it's pretty much a given that it is only bound to get worse once you've begun working together.

Here are some additional factors that small companies need to consider:

- Control or contribution? Compared to big corporations, smaller companies have a hard time controlling open innovation projects. Instead, they should focus on projects where their contribution is truly important and valued. The projects should also fit the overall strategy of the smaller company.

- Big corporations can drain a smaller company. Signs of this include long planning periods, difficulties in identifying and working with the right people, and too much time spent on patent lawyers too early in the process. If these telltale signs appear, a smaller company needs to evaluate whether this will become a drain of valuable resources that could be better spent elsewhere.

- Seeking vs. being found: Smaller companies need to be more active looking around for partners whereas big corporations can focus more on being found and becoming a preferred partner of choice. Companies can look for projects and partners in their own networks (such as customers, suppliers, and partners) or in external networks (such as universities, intermediaries, and consultants).

Some Win-Win Situations Are Better than Others

Here's an undeniable fact of life: Most—if not all—effective ecosystems are initiated and/or controlled by companies that stands to benefit most from them. I do not view this as a negative thing. However, it should especially caution smaller companies when potential partners pitch you a project in which they say everyone will win. Most likely everyone will benefit—if not, then the ecosystem will falter, as no ecosystem can sustain itself if only one partner benefits from it.

We should just not forget that some always win more than others do. Apple is a great example. They walk a thin line as they keep strong control of their ecosystems and reap the most profits. Nevertheless, they still manage to bring in lots of new partners for their initiatives.

A critical task for a smaller company in such a situation is to evaluate whether this is the right setup to join given that you have limited resources and hopefully several options available to you. Keep in mind that not every deal is worth doing. Look for the one where your part of the win-win is most significant for the future of your company.

Starting Small

Now back to the challenge I mentioned in the previous chapter—providing employees with opportunities to build and expand their open innovation capabilities. Corporations that understand the advantages of having smaller companies as part of their open innovation ecosystems can help small companies address this need.

Here are some other resources small companies can tap into to help them build open innovation muscle:

- **Learning with intermediaries:** As the movement toward open innovation has grown, so too has a group of businesses popped up to create innovation marketplaces. These intermediaries help companies of all sizes reach out to talent around the world to solve their innovation challenges. Small companies as well as large can benefit in various ways from using these intermediaries, with one of the benefits being the ability to give employees the chance to build their open innovation skills by being part of a global team. I have much more on this topic in Chapter 9, where I discuss some of the major innovation marketplaces and how your organization can benefit from them.
- **Teaming up with others:** More and more emphasis is being placed these days on clusters, which are geographic concen-

trations of interconnected businesses, suppliers, and associated institutions in a particular field. One of the activities that such clusters can undertake is to provide companies with an opportunity to team with other businesses within the cluster on open innovation efforts. Building on the age-old notion that there is strength in numbers, such clusters can provide training opportunities and other resources that individual companies might not be able to afford on their own.

- Universities, research institutions, and publicly funded projects: Institutions of higher education and research institutes around the world have long partnered with businesses, but this movement has gained even more momentum in recent years. The desire to commercialize innovations developed on campus in order to bring in income is increasingly attractive—and important—to institutions of higher education that have experienced falling revenues from other sources. With universities and research institutes eager to be part of open innovation ecosystems, businesses of all sizes can reach out to these organizations to build relationships that will result in opportunities for employees to engage in open innovation and thus build their skill sets.

- Corporate wild cards: Many larger corporations invite smaller companies into what we can call wild card sessions. These corporations understand the advantages of having smaller companies as part of their open innovation ecosystems and they often invite smaller companies to help with ideation, provide an outside perspective, and provide expertise in a specific area. Giving employees from a small company that is part of a corporation's open innovation ecosystem a chance to participate in such sessions as wild cards exposes these people to the innovation processes and methods used by the larger entity. It also gives both sides a chance to get to know one another so it serves as an effective door opener to a potentially larger, more serious partnership.

- **Consultants:** There is no shortage of consultants and service providers that offer insights and training modules aimed at developing the skills and mindset needed for open innovation.

How Not to Do Open Innovation

In April 2009, I wrote a blog post that picked up lots of interest. It was about Campbell Soup Company and how I viewed them as example of how not to do open innovation. To be fair, they were—and to some extent still are—hardly alone in being a bad example. While their intentions may be good, their execution is poor, and it seems unlikely they will gain the trust of the community they are trying to court with their Ideas for Innovation program. Here are the points I made back in 2009 about the problems I saw with their effort:

- **It's too vague and unfocused.** Campbell says they want "ideas for new products, packaging, marketing, and production technologies that will help us meet the needs of our consumers and customers better, faster, and more completely." That could be almost anything, couldn't it? Why not help your potential external partners and save everybody time by being more specific about what you're looking for? Hopefully, Campbell Soup (and other companies doing this) have set an innovation strategy and know much more specifically what areas they're most interested in pursuing than the catch-all description above implies. By being more specific vague, they would avoid being inundated by useless and energy-wasting ideas.

Turn me on, not off. Campbell says it will take you three to six months to get a reply and if they turn down your idea, you will not receive any explanation of why it has been rejected. Why not try to make it more inviting? I think the reason for both these stipulations is that Campbell is afraid of getting too many submissions, which takes me back to my first point. Bring some focus to the effort and you'll receive ideas that are more on target and that can be reviewed faster and better. And if someone went to the trouble of sending in an idea, you could at least develop some general categories to explain why an idea is turned down.

It reads like an ego trip. A press-release announcing this program said this: "The Ideas for Innovation website is designed to provide an effective way for Campbell to review and evaluate unsolicited ideas by offering people who do not work for the company an easy way to submit ideas." You ask people for their input and yet you design the entire process toward your own needs. Come on! Yes, this is your website and you are in control. But the website talks only about why this is good for Campbell. Why not mention what Campbell can bring to the table to the companies or people interested in working with them and how this can be done?

- No commitment leads to wasted resources and internal resistance. The press release announcing Campbell's Ideas for Innovation initiative mentions that external involvement is a key element to improve their innovation results. If this is a key element for them, then I really wonder why they do not put more effort into it.

Campbell does mention that they have other ways of accessing innovation from other sources. I hope this includes a pure business-to-business version that is much more attentive toward their partners compared to their Ideas for Innovation website. Unfortunately, I have not been able to find any information on this so perhaps they don't. Perhaps Campbell uses intermediaries such as InnoCentive and NineSigma.

The website for this initiative was launched in April 2009. Unfortunately, when I returned to it in May 2011—two full years later—little appeared to have changed. At first, I was amused. Then I got a bit angry. It seems as if Campbell has done nothing at all to improve. The site looks the same and many of the issues I pointed out in my original article still exist. One example is that it still takes three to six months to get a reply—if you get any at all.

I went on to check their corporate site to learn more about their focus on (open) innovation. Nothing showed up.

Then I checked Google. Most people will do online research about a potential partner and their innovation efforts when they are about to embark on an open innovation journey together. It would thus be great to read how the potential partner approaches innovation and how others praise their efforts. You just don't find much of interest when it comes to Campbell and innovation. On the contrary, my original blog post in which I hit on Campbell

Soup pretty hard is one of the first articles to show up when you search for "Campbell Soup" and innovation.

The combination of all this tells me that Campbell is a company to avoid for potential innovation partners and for corporate innovators as well. Campbell simply does not give the impression of being a good place to work if you are a corporate innovator, and I imagine they have a hard time recruiting and retaining strong corporate innovation talent, let alone attracting—and working with—external partners.

Fortunately, everyone else can learn from their bad example; if you are researching a potential partner and they seem to have some of the same flaws that Campbell does in their approach, be wary of what kind of partner they might make. For small companies who have few resources to waste, it is especially important to find the right partner and not one who is as unfocused as Campbell.

Interview #3: Metaio: A Small Company Perspective on Open Innovation

Metaio develops software products for visual interactive solutions combining real and virtual elements. It is a privately held software company founded in 2003 with 65+ employees, based in three locations in Germany, South Korea, and the United States. They have 340+ customers.

A key to Metaio's success is innovation with other partners. In this interview, Jan Schlink, who works in Marketing and Business Development at Metaio, shares some views on how a small company looks at open innovation.

Here's a comment Jay sent me before the interview that I wanted to share; he wrote, "to be honest: I wasn't much aware of the term and circumstances of 'open innovation' (definition-wise) and I hope my answers are not too much in layman's terms.… As far as I understood it, it is more a perspective for huge companies to include SME into the product life cycle at an early stage and share knowledge in order to innovate jointly. Is that correct?"

Jan is very much correct. What I like about this is how a small company actually does open innovation without really knowing the definitions. I think this is very common in many small- and medium-sized companies. They just do it, whereas most big companies engage with open innovation as a clear, deliberate, and strategic choice. Understanding this difference is important when big companies implement open innovation.

Now the interview.…

Why should small companies embrace open innovation?

Schlink: They have the chance to play their specific role in the development of products and services they would otherwise not have had by only marketing their own solution. Especially with consumer products, there are many hurdles an SME can never overcome, but being part of an OEM or an ecosystem can help an SME grow and scale its revenues. It is also very interesting to learn from the bigger partner about processes and Product Lifecycle Management.

What does open innovation mean to small companies? How does this differ from the way big companies view open innovation?

Schlink: The main difference seems to be the market entry strategy for SMEs versus an innovation/technology leadership positioning on the big companies' side. For SMEs it is often an opportunity for their market entry strategy (getting into new markets). The big companies want external know-how to keep their position as the innovation leader or enhance their products with new, unique selling propositions.

What are the benefits of innovating together with others for small companies?

Schlink: In an emerging market, there comes the time when established players enter it with their own solution. Sometimes they have bought a specialist in advance or they developed it themselves. Now, if you are an SME that wants to stay independent, you have to join forces with bigger partners to compete with the jointly developed solution.

In addition, I can mention that marketing and PR get easier with big names and brands. An internal and external multiplier effect kicks in when you are part of an OEM ecosystem; you get highly profiled external feedback on the things you do/did and you get the possibility to create long-term roadmaps and fund R&D for these goals.

What concerns should small companies have about open innovation?

Schlink: Maybe some fear that their knowledge gets "stolen" or that the asymmetric power balance could influence their strategic plans. I think that a good preparation and elaborate contracts together with a comfortable feeling help to prevent being taken by surprise. I bet that big companies with a good "open innovation" image have it easier entering partnerships. Mistrust is a bad starter....

How can small companies prepare themselves for open innovation?

Schlink: They have to be adaptable (daily builds, quality and industry standards); approachable (international, open-minded, soft skills); and self-confident, that is, knowing exactly where their strengths are and that their solutions play a notable role in the joint venture.

Strategically experienced management members are very important. You have to be able to speak the same language as a CTO at a billion-dollar enterprise without sounding cocky.

Which business functions should "own" open innovation at small companies? What kind of people should work with open innovation at small companies?

Schlink: The CEO and/or high-level management should own this if they are technologically experienced, but ideally every unit should be involved with open innovation because projects need to be initialized, created, and marketed.

Can you elaborate a bit on this?

Schlink: Well, let's say there is a management contact between two companies and they agree on an innovation partnership. The project will not be implemented and deployed by these initial persons. Developers will get in touch, same as the marketing or PR department, in order to develop, optimize and, at the end, communicate the project. That's why almost every division should be able to work under these—say, asymmetrical—circumstances.

Open innovation is very much about ecosystems with several partners. Small companies often take the back seat in such setups. Is this a problem?
Schlink: I think this is a matter of time and performance. And I think it's not always bad to be in the back seat. You see all the players.

Small companies often have limited legal resources. What can they do to get better deals and protect their intellectual property?

Schlink: IP management and enough legal resources are important. They should not save the money here! You have a better position in negotiations with a good IP portfolio and a good lawyer who can help you understand complex NDAs and contracts.

Key Chapter Takeaways

- Trust is an essential component of open innovation relationships. As you move toward open innovation, you should begin to look into two questions: (1) What does it take for you to trust others? (2) How do you convince the people in the organizations you want to draw into your open innovation ecosystem to build trust in you and your company and then start forging strong relationships with them?

- Trust is first and foremost established between people and then perhaps between organizations, so it is more important to look at the people side of innovation than the process side.

- The barriers against building trust and strong relationships with stakeholders in your ecosystem may include (1) an internal rather than an external perspective, (2) viewing external partners as someone paid to deliver specific services rather than sources of co-creation and open innovation, (3) being more focused on protecting your own knowledge and intellectual property than opening up and exploring new opportunities, (4) being skeptical about the capabilities of small companies, and (5) being too busy to make it happen.

- Stakeholder management is a critical discipline to master as you try to foster a culture in which trust is a key component and in which resistance to change is minimized so open innovation can thrive.

- You can get an idea of stakeholder management—for both internal and external stakeholders—by thinking in terms of three steps: identification, profiling, and communication.

- Small companies must do their homework to learn as much as possible about the corporation they're about to join forces with, particularly with regard to that organization's history of dealing with smaller companies. This will help you avoid entering into a win-lose relationship in which you get the short end of the deal.

- The more you know about a potential open innovation ally from the get-go, the better prepared you will be understand how your company might be of help to the larger company. This means you will be better prepared to showcase your firm's assets.

- It's easier to start a partnership than it is to end it. But don't be so eager that you don't take time to fully consider whether this is the right partner for you.

- A critical task for smaller companies is to evaluate whether a corporate partner is right for them, given that they have limited resources that a larger company can quickly drain.

- Small companies can provide opportunities for their employees to gain open innovation skills by using innovation intermediaries, teaming up with other businesses in clusters, serving as corporate wild cards, or bringing in consultants to provide training.

7
MAKING IT WORK

Once you've done your due diligence and have decided to proceed with an open innovation partnership, what do you need to do make sure the relationship gets off to a good start and stays on firm footing? In the excitement to begin, it is possible to overlook important groundwork that should be done that will help assure that things go smoothly over time. As usual, always go back to alignment, making sure the goals and outcome are aligned with strategy for both partners. It is also important to establish boundaries, such as deciding up front how any disagreements will be settled as well as being very clear on expectations for the amount and type of resources to be devoted to the project from both sides. This latter point is particularly important to protect the smaller partner who is in the more precarious position regarding the expenditure of staff time and financial resources.

Chris Thoen, former managing director of the Global Open Innovation office at P&G, provided some excellent guidance on this subject in a presentation on "Surviving, and Thriving, in an Open

Innovation Future." While he covered many areas, his main message was to be the partner you're looking for. Here's how he said to go about doing that:

- **It's about actions. But it starts with a mindset that needs to infuse the culture.**
- **Celebrate your partners. They should come back, and tell others.**
- **Don't think in terms of one-off deals.**
- **Unless win-win is the mentality, there are no wins in the long run.**
- **Grow the total pie versus growing your piece of the pie.**
- **Connect partners. Connect suppliers.**
- **Be up front; be transparent.**

This is all important, Chris said, because of what he calls Weedman's law of partnership, which states that the second deal with the same partner takes about half the time and creates double the value.

Jan Bosch, a former vice president of Open Innovation at Intuit, provided additional good advice for the readers of my blog on how to act in the early stages of building relationships. Here's what he wrote:

- **Set expectations up front. We try to be very clear about what potential partners can expect in order to avoid the "false positives" mode. For instance, for entrepreneur day, we promise a yes/no response within forty-eight hours after the event. A "yes" means that we explore the possibility of running a customer trial with the potential partner. We explicitly communicate that even after a successful trial, we still maintain the right to not proceed with that partner, etc.**
- **Avoid confidentiality and IP issues in the early stages. All events and initial contacts are explicitly in "the public domain." No**

NDAs are signed; no promises are made. The goal of initial contacts is to create enough interest from our end to get into deeper discussions that may take place under NDA.

- **Facilitate the building of a community where external innovators may even decide to get together and partner with each other, rather than with us. We've had at least one and probably more cases of where this happened. We are delighted about this and feel it's an important success factor.**

The rest of this chapter is devoted to three case studies that illustrate how different companies are tackling the challenge of building good open innovation partnerships. Each organization's resources and needs differ, of course, so you must find a model that fits your particular situation. However, important lessons can be learned from knowing what others have tried and what has worked—or not worked—for them.

General Mills Town Hall Meetings

As part of the their open innovation strategy, General Mills has developed an effective town hall meeting strategy which helps them efficiently make meaningful connections around the globe with potential partners. Here are insights they shared with me about how to make such a strategy work:

- **When planning your town hall event, select a time when there is already a food or technology industry event happening to ensure that you'll have a large number of relevant people in the area. Then, secure a speaking role for a member of your team at the conference or event to increase the visibility and credibility for your program.**

- Once you decide on a location, partner with a trusted guide in the region who can help you understand the culture and who best to invite to your town hall meeting. For example, you could consider using free resources from trade and export agencies.

- Once you decide who you will invite, encourage potential attendees to visit your website up front to become familiar with your company and the types of technical challenges you're looking to solve.

- During the town hall meeting, explain the heart and mission of your company, what you're looking for, and what opportunities you have for the attendees to partner with you. The meeting also provides an opportunity for attendees to network with each other and chat with you one-on-one for 30 minutes. During the one-on-one meetings, learn about potential partners' core competencies, and what technical challenges they could solve.

- Follow up with the potential partners who were a best fit in terms of core competencies, ability to solve your challenges, and interest in the partnership.

Interview #4: How Existing Partners React to an Open Innovation Strategy

Often, companies already have a variety of strategic partnerships in place, many of them long-standing, prior to deciding to adopt open innovation. If you want these partners to be part of your open innovation ecosystem, you need to carefully strategize how you'll introduce the concept to them. I interviewed Todd Boone, director of Market Development at Psion Teklogix, provider of rugged mobile computers, on this topic.

Last year, Psion took a big step forward with its strategic interpretation of open innovation—called Open Source Mobility (OSM)—when it brought its global partners together for a series of conferences at which they unveiled a new platform designed to give resellers and developer partners the opportunity to co-create with Psion. How existing partners react to open innovation initiatives is crucial and thus I asked a few questions on this:

In which ways did you prepare to bring your partners on board with your open innovation efforts?

Boone: We actually started to bring our partners on board prior to being involved in, or, more accurately, before formally deploying open innovation. By that, I mean that we have had partners developing specific technologies and peripherals on top of our products for many years. Much of the focus is now on formalizing the process to do this and expanding the network further.

Some of the initial process formalization focused on information flows—sharing information that was previously deemed too confidential to go to partners. Further, we have spent a lot of effort improving the "openness" of our devices—both the interfaces into the devices as well as our development tool kits for hardware and software. This gives a more comprehensive platform to our partners to develop on, delivered earlier in the product life cycle. However, there is still much to do in terms of bringing open innovation into the fabric of how we work with partners.

What were the biggest obstacles in getting your partners on board?

Boone: Getting them to understand the full scope of what we are doing. This is understandable because the timing of both change and development within our company dictated the pace and degree to which we could release information. So, it was not until we brought all of our partners together in September that we were able to share the whole message of how we are transforming our processes, our products, and our overall focus to support the adoption of open innovation into our business. Now that they fully understand the focus, they are excited about the direction we are headed as a company.

To our partners, it is critical to communicate the message that open innovation is not just a Psion initiative, but rather an industry phenomenon. Further, they need to understand that opening up our business provides an opportunity for them to claim a larger stake of the value equation by developing unique-to-them solution elements that are of high value for the customer. This is a critical factor in their ability to win deals in a differentiated way.

What were the biggest surprises or learnings when you started to involve your partners in this?

Boone: How positive their reaction was. As we worked on this in the background, there was a degree of concern about how it would resonate with partners. Of course we had shared the full story with a handful of partners for feedback and we had done a lot of general partner research, but it was not until we shared the message with the entire partner community that we could see the true positive reaction.

Also, enabling partners to fundamentally increase their stake in the value equation did not necessarily mean that they would want to. But many are realizing that they can break out of a commoditized market with something unique. The message to customers is that they have access to solutions that better meet their needs. As one partner said, this makes them feel more like "a part of the family." Or, as another stated on our open innovation Web platform, IngenuityWorking.com, we now have a cohesive strategy that "goes beyond a mere reseller agreement."

I think our biggest surprise was the degree to which partners did want to be a part of this. Just as our OSM allows us to differentiate in the market, it also enables our partners to differentiate. Collectively, we are offering a more compelling choice to customers who get to tailor solutions to what they really need—not just what's being offered. So, even though it may require some investment from many of our partners, they understand that they are investing in true differentiation.

Finally, they also see the opportunity for cross-pollination within the partner community. This is critical because despite the tools we have created, not all partners are interested in developing unique technologies, or their focus is in a specific area. However, by augmenting the open innovation initiative with our collaborative community, we have also created the ability for solutions/technologies to be shared across partners, across regions. This expands the market for those that do pursue development initiatives and provides more options for those that don't.

Interview #5: P&G Perspectives on Small Companies and Open Innovation

Here is part of an interview I did with Chris Thoen, formerly at P&G on the intersection of big and small companies in open innovation.

How well informed do you find small companies to be about open innovation?

Thoen: How well a company understands and adopts open innovation isn't linked to its size, but more so to the focus of its leadership, its openness to new ideas, and its commitment to innovation. We've seen small companies deeply engaged in leveraging open innovation to grow their businesses. And others either not committed, or still working to understand what it means and how it can really work for them and their goals.

What have been your key learnings in working with small companies on open innovation projects? What advice

would you give small companies that want to innovate with big companies?

Thoen: A critical learning for us, and a key piece of advice to other companies, is to really understand what a prospective partner needs and expects. For a larger company, it's important that to take the time up front to explain global business needs, time lines, IP issues, and staffing structures. That way, your prospective partner knows what to expect. Larger companies also need to check any possible arrogance at the door—they cannot assume that they know best what a partner might need.

Both sides need to share and really understand each other's pluses and minuses and what's important to each side. That calls for a high level of respect. It's also unwaveringly crucial.

How can big companies improve their ways of innovating together with small companies?

Thoen: To make any collaboration work and work well, you need to ensure you have the right people with the right skills at the table when you *begin* talking, and then work to build and develop the relationship as the project progresses. In many cases, the inclination is to bring in your brightest minds. But that's not the right answer if those people are also arrogant and fixed in their own approaches. Keep those folks in the lab. Instead, bring in the team that is capable of making decisions, but also understands that they are not negotiating deals, they are building relationships.

What are the benefits of innovating with small companies?

Thoen: There are a number of reasons big companies look for opportunities to engage with small companies and entrepreneurs. Key is that they often are at the leading edge on a variety of fronts. They are not bound by the same business limitations as we are in big firms, and don't have a larger, established business to defend and protect. They can be risk takers, and thus their approach and mindset are different. That can be refreshing, inspiring, and a breakthrough for larger firms, and also can be the source for truly disruptive innovation.

What are some early signs of danger/promise when you engage with small companies?

Thoen: When we start talking to a potential innovation partner, we first look to ensure we have an "*integrity match*." Does the other company have the same ethics and values? If not, it will likely surface later in product quality issues and in relationship challenges. Second, is there chemistry? Do we fit, can we collaborate, and do we really like each other? Can we build this relationship into one of real trust? I can't stress enough that this is all about relationships, not just deal negotiations. We have walked away from prospective deals where the innovation was promising, but the fit was simply not right. Looking back, those were clearly the right decisions—for both us and the other company.

What has surprised you the most when working with small companies?

Thoen: I don't know that I'd use the term "surprised," but one of the things we see a lot is companies misunderstanding, and then misrepresenting their own value. Quite often, they either bring to the table an inferiority complex or a superiority complex. Each serves as a disadvantage to them and the prospective partnership.

When they undervalue themselves, they are afraid to fully engage. They worry that they might not have the technical and/or commercial knowledge to be a worthwhile partner.

On the other hand, some enter a meeting thinking they've invented hot water and want to be compensated in platinum. In these cases, discussions are almost always doomed to fail because it's hard to reach really effective collaboration.

It's critical for everyone to be realistic about what they know and have, and present it appropriately to potential partners. Don't under- or over-sell.

Key Chapter Takeaways

- Important groundwork that should be done at the start of an open innovation relationship includes reaching agreement on goals and defining the desired outcome so everyone is confident that the relationship will produce a win-win that is acceptable to both parties. It also includes establishing boundaries, such as deciding up front how any disagreements will be settled as well as being very clear on what the expectations are as far as the resources that will be devoted to the project from both sides.

- Avoid confidentiality and IP issues in the early stages and facilitate the building of a community.

- Consider town hall events to efficiently make meaningful connections around the globe with potential partners. Use these events to explain the heart and mission of your company, what you're looking for, and what opportunities you have for the attendees to partner with you.

- Put in place a strategy to bring existing partners into your open innovation ecosystem.

8
WHY THINGS GO WRONG

It is inevitable that not all open innovation efforts will work as planned. By looking at the reasons why others have failed, you can perhaps be better prepared to avoid such problems. Here are some of the chief reasons I've seen that open innovation efforts go off track:

- **Companies have difficulties making innovation happen internally. Now they hear about open innovation and think of this as an approach that will make everything better. Not true. If you cannot make innovation happen internally, you will have even more difficulties doing this with external partners.**

Obviously, the time to look closely at the innovation track record of your prospective partner is before you join forces for an open innovation effort. But if that wasn't done and you find yourself with a partner that doesn't seem to have its innovation ducks in order, it may be possible to analyze just where the partner is coming up short and see if your organization can help it improve in the aspects of its innovation capabilities that are lacking. This is particularly true for large corporations partnering with small companies; you may very well have knowledge and experience at your disposal that can help the smaller organization strengthen their overall innovation capabilities. Of course, the inverse might also be true in some cases; some small companies may have people who are well-versed in innovation.

If the partnership holds great potential—for example, if your partner has a cutting-edge technology that could be really useful to you—it may be worth your while to help the partner identify the roadblocks it's bumping into and working together to help it overcome them. You might reap great rewards by helping them step up their game. On the other hand, there might be some instances in which you're just better off walking away once you discover that your partner's innovation skills aren't what they need to be for the partnership to succeed.

- **One of the most common reasons open innovation efforts run aground is misuse of idea generation platforms. Many companies start off with idea generation platforms, hoping that external contributors will contribute with great ideas and/or technologies. Most of these platforms do not deliver on the expectations, getting more trash than gold.**

When I wrote about this on my blog, Ellen DiResta, founder of the product development consultancy Synaptics Group, commented, "I've had lots of experience with companies trying to start with idea generation platforms. When they fail to deliver on expectations I usually find the same root cause. They start generating new ideas in a random fashion without first taking the time to define, in market terms, the criteria for a successful idea. They then become overwhelmed with a large number of irrelevant ideas, and then resort to choosing the ones that they know can be implemented."

Deborah Mills-Scofield, an innovation advisor and partner with Glengary LLC, an early-stage venture capital firm, offered these reasons for open innovation failures:

- **Management—"Trying to 'manage' the process vs. 'moderating' the process. While well-intended and rationalized, many open innovation ventures are not done on a level playing field. If it's a large company starting it, they tend to be condescending or paternalistic instead of viewing the other 'innovators' as equals, and this sets a very strong tone."**

- **Trust, or lack thereof —"This is somewhat related to above, but there is a fear of losing control over intellectual property (and hence money), and I've even had clients mention losing people—that the open innovation forum becomes a means of recruiting away their best talent (which in and of itself is a symptom of bigger issues)."**

Additional reasons open innovation efforts crash and burn include

- **Companies copy competitor's initiatives rather than creating their own unique initiatives that match their business reasons for doing open innovation.**

- **Companies fail to make their employees, partners, and customers understand what open innovation means to the company, and they fail to explain the impact of such a new direction to the internal and external stakeholders.**

- **The various organizational units—and in particular the operational ones—are not fully aligned with the innovation initiatives, making it difficult to execute in full on otherwise well-devised initiatives.**

- **Executives fail to understand that issues surrounding the handling of risk and fear of losing control are key to** successful open innovation and thus they don't deal with them head-on.

- Companies put their "best people" in charge of open innovation, failing to recognize that the "best people" who do great by doing things as usual are not necessarily what is needed in order to succeed with open innovation. It is a paradigm shift and you often need different perspectives to succeed.

- Companies focus more on their own gains rather than working toward creating a true win-win scenario.

- Compensation policies and other reward systems don't align with open innovation. If people aren't being rewarded and recognized for taking steps that will move your open innovation effort forward, such as building relationships throughout your ecosystem, then they have no real motivation to adapt new ways of thinking and new approaches to their work.

Corporate Antibodies

Here's another big reason open innovation efforts can fail: they are often killed by corporate antibodies resisting the changes brought by opening up to external partners. These antibodies can exist in both large and small companies. If you're hearing statements such as these, corporate antibodies may be at hard at work, resisting change:

- "We already tried that and couldn't make it work."
- "What we're doing has worked fine for years; there is no need to change."
- "Our current product is still profitable; I don't see why we need to spend money on something new that might not even work out."

- "We already explored that idea years ago but decided against it."
- "If that were a good idea, we'd already have thought of it. After all, we are the experts on this." (Said about an idea coming from the outside.)
- "Let me just play devil's advocate here…."
- "Of course, I support innovation, but I just don't think this is the right time to make a big change. The market isn't ready."

People making such statements may truly believe that they have the company's best interests at heart. Or they may be putting their personal interests ahead of company loyalty. Some people also become antibodies because they don't feel their opinions are given enough weight. Such feelings can cause people to continuously take the negative side or play devil's advocate. The phrase "I hate to bring this up, but…" comes from them a lot, followed by a boatload of negativity.

This is not to say that anyone who questions the need for change or the direction that change is taking is being unnecessarily negative. Sound feedback is needed from many quarters for real innovation to occur. But what I'm talking about is not constructive criticism. Rather, it is the relentless negativity, foot dragging, and throwing up of needless roadblocks that pose a true threat to innovation ever becoming a reality.

Recognizing that corporate antibodies are likely to show up at some point in your innovation process and having strategies in place to deal with them should help you derail some of the people who want to impede change and maintain the status quo. Here are some potential solutions:

- **Make people backers rather than blockers.**

It's never too early to start this. Your initial stakeholder analysis and resulting communications strategy will mean that you're being proactive rather than reactive. By communicating proactively, you can sometimes co-opt the antibodies into the process in a way that satisfies their egos and makes them feel their ideas and authority are being appropriately recognized. You may discover that your proactive efforts weren't enough, but you can continue to communicate to stakeholders that they can play a valuable role in shaping the company's future, including their own destiny. Bring people together to facilitate knowledge sharing and the building of new relationships that broaden everyone's perspectives. Keep people involved in the innovation process.

- **Stay below the radar.**

In some situations, the best choice is to stay below the radar as long as possible. Don't become too interesting too early. This will help you avoid people who want to own the idea or process, or who want to apply standard corporate processes to the project even though this can kill it. This, of course, is more applicable to large companies than to small ones, where everybody knows what's going on and it's hard to keep anything quiet.

- **Have frameworks and processes in place.**

Many internal innovation debacles can partly be avoided by setting internal rules about how to bring innovation projects forward. With a framework and process in place, it becomes easier to move projects forward without having them get hung up in destructive internal warfare.

- **Provide high autonomy.**

In larger organizations, having innovation councils with high autonomy or units with their own assigned budgets and goals are other ways to get around the damage that can be done by corporate antibodies. Such structures help shelter new ideas against situations in which executives are not willing to spend their political capital in supporting innovation or when they believe the change will impact their own career negatively.

Key Chapter Takeaways

- The reasons open innovation partnerships go off track include (1) a lack of capabilities to make innovation happen internally, (2) using idea generation platforms without first defining the criteria for a successful idea, (3) a lack of ability to build trust, and (4) the larger partner trying to manage the process instead of collaborating.

- Open innovation efforts are often killed by corporate antibodies that resist the types of change that being part of an open innovation ecosystem requires.

- Solutions for corporate antibodies include (1) making people backers rather than blockers, (2) staying below the radar, (3) having a framework and processes in place, and (5) providing high autonomy.

9
INNOVATION MARKETPLACES
A MAJOR RESOURCE FOR OPEN INNOVATION

As open innovation becomes more widespread and ecosystems are established, the need increases for innovation marketplaces that can serve as intermediaries to which companies can quickly connect to meet needs and solve problems. Some of these intermediaries will serve niche markets, whereas others will be more general. Some will be set up by companies to meet their specific needs, and others will be set up by third parties that want to position themselves as an interface between companies seeking solutions and the smart people—or companies—with solutions. No matter what they look like, successful intermediaries are—or will become—key players in innovation ecosystems, as they are an important source of knowledge and solutions that should be used early on when companies develop their open innovation initiatives.

Such intermediaries have a role to play with small companies as well as with large ones. One of the challenges small companies need to tackle when they considering engaging in open innovation is getting employees up to speed on the skills that are required to be part of an open innovation team. Intermediaries can help meet this need.

The skills required include collaboration, team building, and communication. Depending on how a small company operates, people are often asked to tackle problem solving on their own rather than engaging in team efforts. So opportunities may be few and far between for building the skills that will make people effective as part of an open innovation team. One solution to this problem could be to ask your people to try working as experts or solution providers on teams at open innovation intermediaries.

One of the better known and established intermediaries is InnoCentive, a third-party innovation marketplace that operates with a prize-based open innovation model. InnoCentive connects companies, academic institutions, the public sector, and nonprofit organizations with a global network of more than 160,000 experts and problem solvers in 175 countries around the world. InnoCentive was originally developed by pharmaceutical giant Eli Lilly as an in-house innovation incubator. An independent organization since 2005, InnoCentive initially saw success within the pharmaceutical marketplace, but it is now active in many other industries, including consumer packaged goods, where companies such as P&G have had success using InnoCentive.

InnoCentive's process works like this: An organization (a seeker) provides a challenge to solvers all over the world who can win cash prizes for solving the problem. More than a third of the solvers have doctorates. Problems have been presented in engineering, computer science, math, chemistry, life sciences, physical sciences, and business.

InnoCentive gets a posting fee and a finder's fee if the problem is solved.

Of course, InnoCentive has competition. Here are few other open innovation marketplaces:

- IdeaConnection bills itself as the world's largest open innovation intermediary. Founded in 2007, the company takes on challenges from large and small companies. They use teams of four to five experts led by facilitators. Their focus is on providing well-researched, in-depth solutions. Awards for problem solvers range from $500 to $20,000 per team member.

- NineSigma provides an extensive global network of scientists, university research departments, and technology incubators to cross-pollinate ideas and provide solutions. Founded in 2000, the company has become a leader in expert-sourcing, offering clients such as GlaxoSmithKline, Phillips, Kraft, Unilever, and Xerox access to its network as well as an extensive database of existing solutions that span all industries and technical disciplines.

- Founded in 1999, yet2.com brings together buyers and sellers of technologies so all parties maximize the return on their investments. Yet2.com focuses on later-stage technologies, rather than on ideas.

- TopCoder bills itself as the world's largest competitive software-development community, with more than 175,000 developers representing more than 200 countries. The community builds software for a wide-ranging client base through a competitive, rigorous, standards-based methodology.

- YourEncore connects companies with retired scientists and engineers, who provide expertise within the life sciences, consumer sciences, food sciences, specialty materials, and

aerospace and defense industries. YourEncore was founded in 2003, with P&G and Eli Lilly as initial clients and the mandate to tap into an underutilized asset: the growing number of retired and veteran scientists.

IdeaConnection goes through a big process in creating its teams, partly by crowdsourcing, by software, by profile, by self-selection, by member and facilitator input, and by IdeaConnection personnel. Teams are facilitated by IdeaConnection moderators. You can also form teams at InnoCentive, but here you are asked to form and run them yourself.

This gives people who are used to working in isolation the opportunity to join with others to solve innovation challenges posed by companies who are willing to offer rewards to have experts from around the world tackle really hard problems.

Time is, of course, a key concern, but perhaps we can think of this as learning by doing and as an alternative to participating in a conference or a training program. Joining with peers around the world who are tackling a challenge means you're learning not just about open innovation but also about dealing with issues of diversity.

Small companies can use the open innovation intermediaries to pose their own challenges and have teams of experts from around the globe work to solve a tough R&D or business problem that may be blocking their success. Several of the open innovation intermediaries work with small companies as well as large. IdeaConnection, for example, has solved challenges for companies as small as one person. Similarly, InnoCentive works with small companies and nonprofits, as well as with major corporations. In contrast, NineSigma, another well-known intermediary, primarily focuses on Fortune 500 corporations.

When you offer up a challenge to one of the intermediaries, you quickly get a world-class team of experts working together to solve your problem or challenge. For example, IdeaConnection has thousands of prescreened experts, many with PhDs, and many with patents to their name. They are professionals in a large variety of fields. InnoCentive's Global Solver network includes people from more than 200 countries. With so many of today's business problems requiring a global perspective to achieve a solution, bringing to bear the brain power of a global community to help you solve a problem has huge potential. In addition, your employees can also participate on the team, allowing them to gain knowledge from these experts and build relationships that may well prove beneficial long after the challenge is solved.

Because the open innovation intermediaries already have all of the systems in place to tackle a challenge, there is no ramp-up time needed. Instead of waiting until you've built your own infrastructure to support open innovation, you can take immediate advantage of the open innovation platform the intermediary has already built.

Before you assume you can't afford to use one of the intermediaries, you should know awards for a theoretical solution at IdeaConnection are usually between $20,000 and $100,000. At InnoCentive, awards have started as low as $5,000 as gone up to $1 million for the most complex challenges. If you have a problem that has stymied your employees and is posing a major barrier to your company's success, using one of the open innovation intermediaries might make financial sense. This is particularly true if it speeds your time to market or helps you beat out a competitor.

Being able to deal with diversity is a key skill that small company employees need to build before venturing into the world of open innovation, but often it's not something they have much of a chance to tackle in their daily work lives. In addition, the exposure to different styles of thinking can be something that such engagement offers that isn't always readily available within a small organization.

There's another potential important benefit here as well. If a small company employee happens to have success with a challenge he or she works on through one of the intermediaries, this will generate enthusiasm for adopting open innovation within your organization. So consider introducing these intermediaries to your employees. Then support their engagement with challenges, knowing that they are learning valuable skills that will, in the long run, support your company's open innovation efforts.

Case Study: InnoCentive and Precyse Technologies

Here's a case study that shows how an intermediary, in this case InnoCentive, can produce results for a company in search of innovative solutions.

Keeping tabs on physical assets in a world that's always on

> It's 8:00 a.m. Do you know where your organization's most valuable physical assets are? You know, the engine inventory you must have staged on the production line just in time to be assembled; those rental vehicles that were returned, washed, and fueled but mysteriously do not appear as ready-for-rent on the online system; or the IV pumps and EKG machines that move so hectically between hospital floors.
>
> Over the past twenty years, organizations in industries like manufacturing, defense, retail, healthcare, and many others have invested billions of dollars in supply-chain automation software such as ERP, leaving the last mile between the data center and the physical world of moving assets un-automated. As a result, assets are always missing

and organizations turn to overstocking, depend on manual labor to register and track assets, and suffer from delays and poor process quality.

How is this problem solved today? By harnessing the power of active RFID (radio-frequency identification) tags—small "cell phones for things" that provide physical assets with a way to communicate their location and status to a network that automatically identifies, locates, and monitors them.

Founded in 2009, Precyse Technologies is one of the emerging leaders in this rapidly growing industry. With N3, the first bidirectional wireless network standard from Precyse, it is possible to deploy real-time asset tracking, sensor networking, and mobile-to-mobile functionality under a single wireless network that ensures five years of battery life for the asset-attached active tag. But sometimes assets are being shipped outside of the facility. By wirelessly switching RFID tags off and on, for example as they pass through the factory gate, it is possible to further extend battery life to much over five years, supporting a stronger return on investment for customers and improving the tags' performance.

Some industry vendors are using short-range HF or UHF passive radio wakeup-calls to switch RFID tags on and off, but this technology delivers only ten feet of range, requiring customers to change their business processes and forcing assets to be tunneled through narrow gates with under 90 percent read rate.

Precyse dreamed of a solution that will allow a process-transparent deployment with a hundred-foot wireless wakeup range. The company understood from the start that taking a traditional approach to R&D could cost well over $1 million. It was money this fast-track growth company preferred to invest in direct marketing and sales. "We assumed a very ambitious goal. Not only was going down the conventional development path prohibitively expensive, it was simply too slow and had no guarantee of success. We wanted to be first-to-market, not the last one to the table," explains Precyse co-founder and CMO Rom Eizenberg. That's when Precyse turned to InnoCentive.

The Challenge: Improve Mobile Device Performance and Battery Life

"We realized that to make the most of our always-on, bidirectional wireless technology, we had to extend battery life beyond the industry standard four- to five-year period," says Eizenberg. "We were looking for a way to wake up active tags as they arrive from transit or enter a customer facility—without draining the battery's energy listening to these wake-up calls. And we wanted to do that without affecting existing customer processes. We had the vision to define the problem, but we didn't have the technical skills in-house to solve it. Fortunately, InnoCentive offered a very appealing alternative.

"Right away, the whole concept of crowdsourcing excited us. Having thousands of the best and brightest scientists and researchers working to help us develop a new technology that could literally change the industry was amazing. The fact that we were only required to pay a Solver should we identify a solution that fit our need was another great advantage. We were basically buying an option on a game-changing intellectual property," Eizenberg adds.

InnoCentive worked together with Precyse to formulate a Challenge, seeking an innovative way to remotely wake up a wireless device without using a battery—at a dramatically longer range than any market available technology would allow. This Challenge was posted to InnoCentive's open innovation marketplace where it was visible to InnoCentive's global Solver network of over 200,000 scientists and engineers in more than 200 countries.

The Solution: An Innovative Way to Harvest Energy from Radio Waves

The InnoCentive community embraced the Precyse Challenge with tremendous enthusiasm and in great numbers. In all, more than 500 InnoCentive Solvers from 64 countries contributed their ideas and technical expertise to help solve the Precyse Challenge. Within 90 days, Precyse received 33 relevant submissions—several of which offered novel, innovative solutions.

"I was impressed by the quality of the ideas and the diversity of the technical approaches to solving our Challenge," says Michael Braiman, Precyse co-founder and CTO. "InnoCentive gave us the freedom to sort through a wide number of Solver solutions and pick the best of breed."

After narrowing the submissions down to three, Precyse selected a solution that was truly breakthrough, both in its approach and execution. According to Eizenberg, "Our proprietary passive wakeup technology allows our new Smart Agent tags to harvest energy from radio waves 100 feet away from a gate transponder. This new technology will allow us to extend battery life much beyond the currently available five years and gives us the flexibility we need to add new features as our technology evolves."

The Results: Faster to Market, More Power to Precyse

"We now have a product that delivers a more efficient approach to energy saving, lowers the cost of tracking physical assets, and improves workflow for many of our customers," says Eizenberg. "Our technology does for 'things' what the cell phone and the Internet did for people. It allows inventories in a supply chain or products on a manufacturing line to talk to one another, to people, or to computers more efficiently and with a faster ROU than ever possible before. Continuing to innovate new product capabilities on top of core product development requires you to scale rapidly while maintaining a lean and flexible structure. Tapping into the crowdsourcing power of InnoCentive's open innovation marketplace presented us with a cost-efficient way to scale while keeping R&D costs in check."

Key Chapter Takeaways

- Innovation marketplaces can help small companies meet the challenge of getting employees up to speed on the skills that are required to be part of an open innovation team.

- Small companies can also use the open innovation intermediaries to pose their own challenges and have teams of experts from around the globe work to solve a tough R&D or business problem that may be blocking their success.

- If a small company employee happens to have success with a challenge he or she works on through one of the intermediaries, this will generate enthusiasm for adopting open innovation within the organization.

- InnoCentive, NineSigma, IdeaConnection, TopCoder, and YourEncore each offer a different model; you should explore them all to see which one suits your needs best.

10
IPR AND OPEN INNOVATION

One of the first concerns that pops into many people's minds when they first learn about the concept of open innovation is, "What about our intellectual property? How can we protect that if we're working in an open innovation environment?" Interestingly, I have seen a shift over the last couple of years as more companies have begun to find success with open innovation. Questions about IP have become fewer. So while this is still a very important issue, I have noticed that it is no longer one of the main topics in the presentations you hear from corporate people presenting at conferences. The reason for is that the companies that are leading the way in open innovation understand that business needs to come before legal issues, and they have gained experience in making this happen.

The legal departments at these leading companies have been turned around. They now think offense rather than defense. "Yes, there will be challenges with this partnership, but there are also opportunities.

Let's find a way to work around the challenges." Another thing is that they have developed a range of simple documents and approaches that better allows them to engage with potential innovation partners. These simple approaches allow them to focus on opportunities before they need to look into the more complex legal issues.

A news editor for *Wired* magazine interviewed me on this topic. Here is an excerpt from the interview:

Wired: Do you think the current legal system is able to deal with the idea of open innovation?

Yes. Companies that are successful with open innovation have no need to change the legal system. They have just found ways to circumvent some issues that could hinder open innovation—– and these issues are almost all within their own organizations. The keyword is "mindset." Are the legal people thinking offense or defense? Can they see the opportunities or just the challenges? Business-minded lawyers are needed for this.

Wired: How do you balance the idea of intellectual property and open innovation? Are they opposed to each other?

No. The challenge is to make them work together. You can argue that companies only do real open innovation when they create new IP. It is easy enough to decide who brings what to the table and how to compensate for this. The real challenge starts when you co-create new IP. This goes back to my reply to the first question.

Wired: What is the biggest challenge for lawyers/companies when it comes to trying to become more open and share knowledge?

In my view, they seem to be trained—and some are very well trained—to minimize risks rather than to see opportunities. This is the biggest challenge.

Wired: How do companies move from being very closed and secretive to being open and collaborative?

This is a change process. You cannot just do this overnight. It takes years in which you start out small, get some early successes, and then convince more people within your company and relevant ecosystems that the open and collaborative approach brings more benefits than what comes from being closed and secretive. It is also important to be able to show benefits not only for the company, but for the individual person as well.

Legal Aspects of Open Innovation for Small Companies: The P&G View

I asked Chris Thoen, who at the time was managing director of the Global Open Innovation office at P&G, this question: "Small companies often have limited legal resources. What can they do to get better deals and protect their intellectual property?" Here is his reply:

> "This is an area of critical importance—to both parties involved. Unless everyone is comfortable with and clearly understands the goals, parameters, and expectations of a project and of both parties, the relationship will never grow into the deep and trusted collaboration needed to deliver meaningful, maximum results.
>
> "Often, small companies do not have the same access to deep and experienced legal teams of their larger partners. Because of this, and the significant cost associated with hiring outside counsel, they will let the larger company take the lead, which usually means the first step is drafting up a

full contract. The result can be a very lengthy full-fledged document complete with legal jargon and sentences that run over three pages that no one really understands. Invariably, the smaller company will still have to pay an attorney substantial fees to first translate everything and then even more to be involved in finalizing all the terms.

"While a full contract will be needed eventually, it's not a good way to start working together. What we've seen work exceptionally well is for the companies to begin by creating a simple, straightforward one- to two-page letter of understanding that outlines, in simple language everyone can understand, the overall working principles for the partnership. This will help flush out whether there is alignment as to key legal and technical needs for each party to ensure a 'meeting of the minds' in advance of time and money being put into the legalese of a contract.

"Small companies should ensure that the letter clearly addresses

- How the partnership will deal with IP generated by ongoing work and with 'background' IP that is brought into the relationship. Will there be exclusivity?
- What happens in the event of success and more importantly, what happens if the joint work is not successful? For example, can either company take their work, and what portion of it, and try again with another partner?"

5 Biggest IP Legal Mistakes Small Companies Make When Working with Large Companies

Jackie Hutter, an IP strategist with The Hutter Group, LLC, believes it's possible for big and small companies to work together in open innovation. But to do so they must avoid these IP mistakes that Jackie has identified. Here's what Jackie has to say on the topic on her blog[1]:

> This is a subject very near and dear to my heart, as I am currently moonlighting as general counsel of a start-up energy company that is moving toward licensing our technology into large companies. Also, as a senior IP lawyer at a multi-national consumer products company, I was on the other side of such deals on more occasions than I can count. Prior to that, I was a law firm partner representing large and small corporations in patents and licensing issues, and in doing so, I now realize that I killed more deals than I ever facilitated—a situation that is more typical of law firm lawyers than it should be, unfortunately.
>
> In view of this multifaceted experience, I present this list of the five most common mistakes [small] companies make when working with large companies in open innovation.
>
> 1. Thinking you have all the answers for the large company's problems: As a small company, you often have only have a single idea or technology and you quite properly focus your attention in this direction. This can be damaging to your ability to complete an open innovation deal with a big company, however. The large company may not care about what you see as the value of your technology because they

are the experts in their products and customers. Indeed, you may be wholly wrong about why the large company is interested in speaking to you. If you want to sell or license your technology to a large company, your best bet is to focus on the specific technology aspects, and leave the business and customer issues to the other side, at least at the early stages of discussion.

2. Bringing the wrong lawyer to the table: Very often, small companies assume the lawyer who handles their intellectual property issues is the appropriate person to bring to a conversation with the big company. However, the legal skills a small company needs to obtain its patent rights are very different than what are needed to get a deal done. While protection of your small company's IP should be paramount in any dealings with the large company, putting up complicated restrictions about the use and ownership of your IP even before you know a deal is likely to happen, which is the natural inclination of most IP attorneys, can often end up in the other party walking away before a deal is even underway.

I have found the best lawyers to negotiate deals in the open innovation context are business-focused attorneys, who tend to be people who have served a stint in the corporate world and who might even have little experience with high-end IP issues. It can be tough to find someone with these credentials, however. As an alternative, I like to work with licensing experts, most of whom have successfully closed more deals in a year than many law firm IP lawyers see in their entire careers. These licensing experts are frequently not lawyers, but they have negotiated enough agreements to be very good at spotting the legal issues and, in this regard, they often do a better job than lawyers.

3. Putting the legal issues ahead of the business issues: Even if you hire a business-focused attorney or licensing consultant to deal with the large company in an open innovation context, deals can still go awry when legal issues are discussed before a business agreement is made in principle. When working with small companies seeking to license their technology, I typically tell them that the last person they want in the initial discussions is a lawyer. (Interestingly, this is the opposite advice I gave when I was a lawyer billing my clients on an hourly basis.)

I can coach a competent businessperson about the basic legal issues they need to know while building a term sheet with their counterparts at the large company, but as a lawyer, I can never fully understand their business and financial details at the level needed to sufficiently spot these issues. If the business parties have established a good working relationship, then it should be fairly easy to renegotiate a term sheet if it is later found out that the business agreement has resulted in legal issues arising. Of course, a deal could still get done if the lawyers lead the process, but I can almost guarantee you that the process will not only cost you significantly more money, but also more aggravation if you put your attorney out in front.

4. Not understanding the economics and low probability of success of product development: Often, small companies believe that licensing to a large company means that a big payoff will be forthcoming immediately and for a long period in the future. Hypothetically, if I was to license my start-up technology to a large company today, I should expect that a product would not be in the market for twelve, eighteen, or twenty-four months or more, and that it will cost the large company hundreds of thousands or often

more than a million dollars to introduce a product into the market, depending on the specific product. Moreover, the failure rate of new products is well above 75 percent, so it can be expected that even the best technology has a better than even chance of failure.

Of course, by the time the small company enters into negotiations with the large company, the small company has borne all the risk; however, moving forward, the large company will now acquire substantial risk. Rather than seek a big payout on the front end, I strongly suggest developing a licensing arrangement directed toward risk sharing by both parties. Certainly, the large company likely has a greater ability to bear risk than a small company, but when the small company shows it is willing to have "skin in the game," deals are more likely to get done.

5. Assuming the large company intends to "stick it to you:" The prevailing view of the dynamics of when a small company seeks to license or sell its technology to a large company is that the latter wishes to pay as little as possible or, even better, to crush the little guy so they don't have to pay at all. This perspective is bolstered by Hollywood and the business reporters who like to find situations involving drama, like the theft of patent rights or the like.

In reality, however, large companies engaging in open innovation today have made the decision that they need partners to be successful in their product development efforts. The decision to go outside of one's own company is one that signifies a collaborative environment at a large corporation, and it is less likely that a company that has truly embraced open innovation will be motivated to "crush" or otherwise debilitate their partner. Why would they? If the small company thrives, the large company will likely end up making even more money.

The fact that identifying and developing a new partner costs much money and opportunity costs also cannot be ignored when considering the motivations and actions of the large company. This is not to say that there will not be disagreements in the dealings between the large and small companies; lack of alignment is inevitable in any situation involving collaboration. However, in today's open innovation environment, I think it is wrong for a small company to question the large company's motives when a disagreement arises. This is a further reason for business people at the respective companies to develop a relationship before bringing in the lawyers, because brewing disagreements can be more easily settled by people who understand their partner's business goals before a full-blown conflict arises.

The above list is a modest attempt to identify the issues that I have often seen arising when large and small companies deal with each other. Of course, lots more can go wrong in open innovation dealings than I have laid out in this post. The best advice I can give to small companies that are seeking to sell or license their technology to a large company is akin to the advice many long-term married couples often give to others who want to know what their "secret" is: become friends first, try to understand what the other party wants out of the deal and don't be afraid to talk out your disagreements. In the end, all successful business partnerships center on building relationships and strong communication, and open innovation is no different.

Key Chapter Takeaways

- Concerns about intellectual property issues in open innovation have been decreasing because the companies that are leading the way in open innovation understand that business needs to come before legal issues and they have gained experience in making this happen. The legal departments of these leaders now think offense rather than defense.

- Common mistakes that small companies make when working with large companies in open innovation include (1) thinking they have all the answers to the large company's problems, (2) bringing IP lawyers instead of business-focused lawyers to the table, (3) putting legal issues ahead of business issues, (4) not understanding the economics and low probability of success of product development, and (5) assuming the larger company has bad intentions.

1 Jackie Hutter, "Open Innovation Insights: 5 Biggest IP Legal Mistakes Small Companies Make When Working with Large Companies," *IP Asset Maximizer Blog*, February 3, 2011, http://ipasset-maximizerblog.com/?p=1247.

11
USING SOCIAL MEDIA TOOLS

Social media, by most definitions, blend technology and social interaction for the co-creation of value. This fits well into what I see as the underlying concept of the intersection between social media tools and open innovation: it is about how we can involve various stakeholders in creating better innovation outcomes. It is worth noting that almost all of the intermediary companies operate via social media. Also, the extensive networking that is a prerequisite of open innovation makes it a natural for linking people together through social media. This chapter looks at how social media can assist you in building better partnerships and stronger open innovation efforts.

The more I work on the intersection of social media tools and open innovation, it becomes clearer to me that companies need a multi-target approach to efforts in this intersection.

One reason is that most open innovation efforts lie in three different circles. There is often a significant overlap, but we still need to look at each of them separately. These circles are:

- **The innovation community: Here we have thought leaders, academics, consultants, service providers, and corporate voices on innovation. This community can bring credibility to the innovation capabilities of a company, and this credibility can turn out to be very important in the next circle, the innovation ecosystem.**

- **The innovation ecosystem: This is the most important circle for companies serious about open innovation, as this is where things actually happen. The ecosystem includes the partners (primarily other companies, institutions, and universities) that help create the innovation output. It is very much about business-to-business.**

Today, very few companies are limited to just one choice when it comes to picking innovation partners. This takes us back to the credibility that can be earned in the innovation community. Let's say a potential partner Googles your company and discovers that your company is mentioned in blogs or articles and that you or your colleagues talk at conferences. This will put you in a better position than other potential partners not having this credibility. Of course, this goes both ways; you also want to check on the credibility of potential partners.

- **Customers and users: Some companies and industries more than others need to pay serious attention to the third circle, which is made up of their own customers and users as well as those of their partners in the innovation ecosystem. This is more about business-to-consumer than the business-to-business focus we see in the innovation ecosystem.**

The importance of each circle varies from company to company and so does the overlap between the circles. Companies need to find out how this works for them and since the circles are different, they also need to apply different social media tools to them.

How Social Media Can Help

Social media offers two key aspects that support open innovation: (1) communication of relevant messages and (2) collaboration, including sharing of ideas and solutions. Here are some benefits of engaging in social media in your open innovation efforts:

- **Better access to and interaction with stakeholders**
- **Feedback loop on ideas and projects**
- **Business intelligence**
- **Marketing and promotion of projects and innovation outcomes**
- **Thought leadership position**
- **Training on innovation skills**

Many skeptics do not see much value in using social media for open innovation today. This is fair enough as it is indeed hard to find good cases and evidence on such efforts, but please remember that we are still in the very early phases on this intersection of social media tools and open innovation. I urge everyone to look two years ahead. This is where things will really start to fall into place as we all get more experience with tools and services that continue to develop at a fast pace and in directions that are hard to foresee.

I ask you to be the visionary company in your industry. Expose your employees and your external stakeholders to social media and learn as you go. Yes, there will be initiatives that do not work, but you will adapt and the experiences gained can bring competitive advantages in the short-, mid-, and long-term.

What Works

Andreas Kaplan and Michael Haenlein, two academic researchers at ESCP Europe, state that there are six different types of social media: collaborative projects, blogs and microblogs, content communities, social networking sites, virtual game worlds, and virtual communities.[1]

Within these categories, I find these tools to be the most relevant for open innovation efforts:

- **LinkedIn: Knowledge is the key element to innovation, and LinkedIn is a great tool for identifying people with knowledge. This works especially well if you upgrade to a business account. It is also possible to get good replies if you start a discussion in the LinkedIn groups, but there is unfortunately also too much noise (spam) in these groups.**

- **Twitter: First, a heads-up. Twitter is practically useless unless you use an application such as TweetDeck, which allows you to filter through the crazy stream of content. Once this is up and running, you have a great business intelligence tool that allows you to track topics that are of interest to you. Twitter can also be used to broadcast your messages, although you do need several thousand followers—and relevant content—to see real results from this.**

- **Communities: Open innovation has many channels for business opportunities—virtually as well as physically. There is a growing need for companies to have a strong destination site, and I believe we will see a shift in which this has to look more like a community rather than pure needs/assets sites such as P&G's Connect + Develop. Communities will have**

a mix of content, business functions such as needs/assets listings, and social networking features.

I do not include Facebook as this is a personal tool for me. However, in industries with lots of consumer interaction, Facebook might also be a useful tool for open innovation efforts. In addition, while I believe LinkedIn and Twitter are the most relevant social media tools for open innovation at this point, you should certainly consider YouTube, blogs, SlideShare, Quora, and others as well. This also includes service providers and intermediaries such as InnoCentive, NineSigma, Yet2.com, Spigit, and HYPE Innovation as they have begun building social elements into their offerings. The key is to look at and work with many different tools and build a "system" in which you capture value out of all these tools at the same time.

An important element of capturing value is to be able to direct interested people and companies to the "destination site"—the corporate site where you not only have a vibrant community where stakeholders can learn from each other and co-create value but also a place where you can see the needs and assets of the host company. Once you start getting conversations going and content begins to develop, you need to start up processes that allow you to extract value out of this information and interaction. No doubt, information overload will occur, but I view this as a luxury problem compared to having no information overload at all.

A word of caution: I have seen several companies try to do social media the easy way. They thought they could set up a LinkedIn group and be done. For example, a major global company set up a LinkedIn group around their industry focus. They managed to collect over seven hundred members but activity in the group slowed to a trickle, partly because they set up very rigid processes; for example, most of what the company could share with the LinkedIn group had to be approved by the communications department. Fitting the management of a social media effort into an already full workload is hard enough, let alone having multiple hoops to jump through before you can do anything.

Action Plan

Strive to become the preferred partner of choice within your innovation ecosystems. This status can be achieved by facilitating a community that is acknowledged as a key innovation resource by relevant innovation partners in business areas that are important to your company. The status can be build on three elements: (1) the destination site, which is a company-controlled platform/website that functions as the hub of the community and shows your innovation needs/assets; (2) thought leadership activities through social media channels; and (3) physical events that allow the community members to meet face-to-face. Neither physical nor virtual activities can stand by themselves; they need to be integrated with each other.

Here's your action plan:

Short-term actions (0–3 months)

- **Build an understanding of your strengths and weaknesses related to achieving the vision.**
- **Develop a better understanding of social media tools and how they can be used.**
- **Identify internal and external members and form a team that can help the company become more visible through social media tools.**
- **Identify key target people for the community.**
- **Develop a process for creating relevant content and develop a strategy on how to share this content.**
- **Develop a strong storyline for upcoming communication efforts (communication will be a key element internally as well as externally).**

- Identify the key obstacles to developing a company-controlled platform/website and begin lobbying in order to overcome these obstacles (most often internally).
- Identify and secure speaking slots at relevant conferences.
- Identify key influencers within the innovation community and within the innovation ecosystems, and develop a strategy on how to "influence" them.

Mid-term actions (3–12 months)

Besides a continued development of the short-term actions, you should

- Launch a beta version of a company-controlled platform/website.
- Develop a format for physical events and begin hosting events.
- Educate senior executives on open innovation in general and the intersection of social media tools and open innovation in particular. (Goal: have one or more senior executives join the team.)

Long-term actions (12–24 months)

Besides a continued development of the above actions, you should launch a full-feature company-controlled platform/website and continuously explore ways of developing this into a strong community.

What It Takes to Make a Community Work

The first question people ask when the topic of building a community comes up is "How can I make this work?" I have witnessed several initiatives and run my own groups and networks in both the virtual and physical world. Based on this experience, I believe these five elements are key to success:

1. A Genuine Need: Many community owners or initiators take it granted that their intended target group has a need for this group or community. This is not always true, and you need to remember that you are asking for the most precious thing potential participants have—time. You need to find some good reasons for "bothering" them with yet another community or group.

2. Value: If there is a genuine need for connecting and enabling stakeholders within a community or ecosystem, you still have to point out the specific value. One little example could be to help provide an overview of what is going on in the given ecosystem. Information overload is a big problem, leaving opportunities for community owners to help solve this.

3. People: You need to have people building the community who really believe in the need for such a community and see its potential for value creation. They must listen to the stakeholders in order to build the features needed, and they must act as facilitators, helping the stakeholders get to know each other and helping them to get value from the community.

4. Communication: If you are selling a vision, as you do when you launch such a community, you need to have a strong communication plan. What stories can be used to recruit the right members and make the current users even more active? Remember that communication on your open innovation messages is just as important as communicating on products and services.

5. Persistence: It is very difficult to create successful communities that are worthwhile to the stakeholders while also bringing innovation or business opportunities to the owners. It requires lots of experimentation to find the right model, and this requires lots of persistence—and time.

This is also caused by the fact that a well-run community succeeds not only by delivering value for the stakeholders, but also because of the trust and respect it has gained. This takes time.

Are There Any Negatives About Using Social Media for Open Innovation?

Here's a quote from a *USA Today* article that tackled the downside of social media—information overload: "People are drowning in a deluge of data. Corporate users received about 110 messages a day in 2010, says market researcher Radicati Group. There are 110 million tweets a day, Twitter says. Researcher Basex has pegged business productivity losses due to the 'cost of unnecessary interruptions' at $650 billion in 2007."[2]

The concerns about the hardship of keeping up with a constant barrage of tweets, texts, blogs and instant messages are real enough. However, social media tools will be key for making innovation happen in the future so we need to find ways of coping with this.

My advice is quite simple:

- **Reason: Find out what you want to get out of the tools. Is it to gain or share insights or both? If you put some thought into this, you are better prepared when you look into the many options.**
- **Focus: Once you know what you would like to get out of social media tools, you need to focus. There are so many options and you cannot cover them all. Focus, focus, focus.**
- **Filter: This ties back to focus. Tools such as TweetDeck (a Twitter application) are very good at helping you filter**

the information flow and thus keep your focus. Find and use them.

Here's another possible concern about social media and innovation: What happens with innovation if executives and managers begin to fear social media tools like Twitter, YouTube, Facebook—and perhaps even LinkedIn?

I wonder how big this issue really is after having read about a survey from Chef.se, a Swedish website for executives and managers. It showed that an increasing number of executives and managers are afraid of ending up—involuntary—on the above sites. Furthermore, 18 percent said they know of incidents where employees had posted inappropriate or false information about their companies using social media tools.

I have always sensed some reluctance toward the new social media tools from the upper ranks. Unfortunately, it becomes somewhat understandable with the above information, although I still believe benefits such as faster communication and better reach clearly outweigh drawbacks that tend to evolve regarding loss of control issues.

Perhaps this is not such a big thing today, but it might grow in the coming years. I got to thinking of a recent comment made to a blog post by Michael Fälling Soerensen, CEO of Nosco, an idea management software company: "Today, teens take pride in sharing and collaborating as much as possible—and are constantly seeking to expand their network—all vital factors in getting an innovative culture up and running…."

I think this is quite true and it even extends up to people in the early twenties. This will have an impact on innovation culture in many companies as this generation enters the workforce in full strength. Their mindset fits perfectly with elements needed for a more open and externally focused innovation culture so we can expect to see interesting—and different—things from this work group. That is, if the executives and managers see this as an opportunity and release the potential rather than treating it as a threat and as a sign that they have to cede control.

Twitter and Innovation: A Great Example from Psion

Last year I organized a Twitter chat with and for Psion, the maker of rugged mobile computers.

The topic was "Open Innovation: A View from the Top." Psion made a strong commitment by having three top executives participate, including their CEO. Some outcomes included

- **Many tweets and retweets (spreading the word about Psion)**
- **More followers to the Psion team**
- **Blog posts written by others about the chat (You can find an example at http://wp.me/pXewP-dR.)**

Most importantly, this gave the Psion team lots of inspiration on what you can actually do with Twitter to promote your innovation capabilities and interact with current and future stakeholders in your ecosystems.

Psion's Director of Market Development Todd Boone is heavily involved in implementing Psion's open innovation strategy and this is how he reflected on the Twitter chat:

"As Psion evolves its business, it's imperative to begin the shift from strategic ideas to actual practice—operationally in terms of how we approach the business actively marketing the thought leadership aspect of our transition.

The expectation is to demonstrate this thought leadership through an increasing number of initiatives—both events and virtual transactions that enable us to target a broader audience then we would traditionally communicate with.

This Twitter chat was an ideal first step toward doing this—instead of constraining ourselves to our typical customers, partners, and even competitors, we saw this as an opportunity to instead reach out to a broad group focused on the same open innovation philosophies that we are.

It is still early days, but our expectation is to expand our network and knowledge base dramatically over the coming months.

I like how Psion is curious and brave enough to try out new things, and I really like how Psion is willing to share its experiences with others. All of these characteristics are important to making open innovation work.

I've done other Twitter chats on open innovation with corporate executives, including chats with P&G and General Mills and with open innovation experts such as Henry Chesbrough. This has proven to be quite useful for sharing content and ideas and for "meeting" interesting people. Consider how you could organize such chats with members of your own open innovation ecosystem to share insights. You should also look into existing chats such as InnoChat (www.innochat.com), which happens every Thursday at noon (Eastern US time).

What's Next?

The pace of evolution in social media is astonishing, so before leaving this topic we need to consider where things are headed. If we look first at companies who are doing a great job now, we can look at Intuit (http://www.intuitcollaboratory.com) and P&G (www.pgconnectdevelop.com). Intuit does well at merging virtual and physical activities and P&G excels at explaining their needs and available assets and with opening their site to people around the globe through having the site available in different languages.

Despite the good work these companies are doing, I think their sites (and others) need to evolve into next generation sites that include more community features. Three companies that have already moved in this direction include GE (challenge.ecomagination.com), Psion (www.ingenuityworking.com), and SAP (www.sdn.sap.com/irj/sdn/coil). What I like about these initiatives is that they are setup for interactions and conversations. They help create forums where everyone can learn from each other while giving the host companies some interesting opportunities to capitalize on all these interactions—if they manage to set up systems and processes to extract this value. They are not just about showing needs and assets—they go social to create a community. If you want to be the partner of choice in your ecosystem, this is the direction to pursue.

This intersection between open innovation and social media tools is interesting, as it brings new opportunities. Furthermore, there is a huge lack of knowledge and experience on how to do this, so companies that are willing to experiment to learn and develop have the potential to reap important advantages over competitors who are slower to adapt these important new open innovation tools. I really look forward to seeing how companies will do this and I plan to capture such insights and learnings and use them or my upcoming book on the intersection of open innovation and social media.

Key Chapter Takeaways

- Companies need a multi-target approach to using social media in their open innovation efforts because these efforts lie in three different yet overlapping circles: (1) the innovation community, (2) the innovation ecosystem, and (3) customers and users.

- Social media offers two key aspects that support open innovation: (1) communication of relevant messages and (2) collaboration, including sharing of ideas and solutions.

- LinkedIn, Twitter, and communities built around destination sites created by companies are the three most useful social media tools for open innovation today. But this is a constantly evolving field so you need to stay alert for new developments.

- To help build a status as a preferred partner of choice within your ecosystem, use three elements: (1) the destination site, which is a company-controlled platform/website that functions as the hub of the community and shows your innovation needs/assets, (2) thought leadership activities through social media channels, and (3) physical events that allow the community members to meet face-to-face.

- The five elements needed to make an open innovation community successful are (1) a genuine need, (2) value, (3) people, (4) communication, and (5) persistence.

1 Andreas M. Kaplan and Michael Haenlein, "Users of the World, Unite! The Challenges and Opportunities of Social Media," *Business Horizons* 53, no. 1 (January–February 2010): 59–68.

2 Jon Swartz, "Social Media Users Grapple with Information Overload," *USA Today* (February 2, 201(1), http://www.usatoday.com/tech/news/2011-02-01-tech-overload_N.htm.

12
KEY CHAPTER TAKEAWAYS RECAP

Chapter 1: Definitions

Key Chapter Takeaways

- Open innovation is about bridging internal and external resources and acting on those opportunities. This contrasts with closed innovation, where you do not attempt to assimilate input from outside sources into the innovation process, and you avoid having to share intellectual property or profits with any outside source.

- Open innovation should take place throughout your innovation process, not just in the early phases during what we call the front end of innovation. You will miss out on the full potential of open innovation if you more or less deliberately shut down to external resources later in the process.

- With open innovation comes the need to create value networks—or ecosystems—that include all the potential categories of external sources that can support your innovation effort. This may include customers, suppliers, academic institutions, and even competitors.

- Before jumping into open innovation, determine why open innovation is relevant to your company, its present situation, and its mission and vision.

- Each company needs to find its own approach to open innovation, one that matches its objectives, capabilities and resources.

Chapter 2: The Benefits of Open Innovation

Key Chapter Takeaways

- Just because small companies don't see themselves as innovators doesn't mean they are not innovators. In fact, they use the market as their innovation lab as they constantly respond to the needs of their customers.

- According to Intuit, necessity, opportunity, and ingenuity drive innovation in small businesses.

- Open innovation can speed the development and market launch of new products and services, bring more diversity to innovation, improve the success rate of new products and services, and diversify risks and share both market and technological uncertainties of innovation.

- Open innovation addresses the issues posed by rapidly advancing science and technology and opens up your company to talent from around the world, enabling you to fill gaps in expertise.

- The pursuit of newness through real innovation is what separates the companies that thrive from the ones that languish in the doldrums or even fail altogether.

- There is a strong possibility that the best people and the best ideas are to be found outside your organization. Failure to recognize this can hold a company back from realizing its full potential.

- Before embarking on a journey toward open innovation, you must explore what open innovation will do to your business model, understand how your organization will change to accommodate open innovation, and determine whether you and your organization really understand open innovation and what it takes.

Chapter 3: When Big Companies Meet Small Companies in the Open Innovation Efforts

Key Chapter Takeaways

- Small companies bring these advantages to corporations that are striving to become preferred partners of choice within innovation ecosystems: (1) Small companies often are at the leading edge of breakthrough or disruptive innovation. (2) Small companies can take risks that large companies can't afford to take because the bigger entities have to protect and defend their established core business operations. (3) Smaller companies are often closer to the markets they serve than large corporations are to their markets. (4) The approach and mindset of those operating in

Key Chapter Takeaways Recap

  small businesses can provide a breath of fresh air to large corporations.

- Open innovation is a paradigm shift in which companies must become much better at combining internal and external resources in their innovation process and act on the opportunities this creates.

- If you want to bring in external partners to your innovation process, these partners expect you to have your own house in order. Adapting open innovation will not eliminate your internal innovation problems; you need to solve those first.

- When you begin to innovate with partners, you will see that these partners either focus on their own needs—and then innovation will definitely fail—or you will see that they come together and funnel their resources toward a market need. If the latter happens, then you have a great chance to succeed with innovation.

- When small and large companies intersect for open innovation, they need to overcome differences that may include speed of decision making, attitudes toward risk, willingness to develop new rules instead of following the old ones, allocation of resources, definitions of innovation, and varying processes or a lack of processes. They also need to be clear about each other's business model and who is in charge.

- As you work with external partners, you are exposed to other ways of getting things done. You bring diverse thinking into the organization. This can make you consider whether your current practices are good enough, or whether you have to adjust these or perhaps even develop new next practices for your organization.

- Large companies that hope to succeed in becoming partners of choice must do their best not to use their size to get their way all the time in open innovation partnerships.

Chapter 4: Getting Your Organization Ready for Open Innovation

Key Chapter Takeaways

- If your strategic plan is several years old, before trying to answer questions about your innovation strategy and your open innovation, you'll need to update your plan to make sure your strategy still fits with shifts that have occurred in your marketplace as a result of competitive or economic changes or technological advances.

- Finding a definition for open innovation that meets your company's individual needs, resources and market situation is essential. Once you've figured out why you need open innovation and how you're going to define it for your particular company, it becomes easier to work out a strategy and implement it.

- To have your open innovation strategy succeed, you must put in place these five elements: (1) stakeholder analysis, (2) communication strategy, (3) common language, (4) appreciation of employees, and (5) networked innovation culture.

- Keep open innovation in the forefront of employees' minds by (1) making it relevant and showing success, (2) promoting it, and (3) making it easy and making it stick.

- Key components of a good networking culture include (1) clear strategic reasons why employees need to develop and nurture internal and external relationships; (2) an understanding of the types of networks you hope to build to support your innovation efforts; (3) leaders who show a genuine and highly visible commitment to networking, who walk the walk, not just talk the talk, and who share examples of their networking experiences whenever possible; (4) networking initiatives that mesh closely with your corporate culture; and (5) both virtual and face-to-face networking opportunities.

- People must be given the time and means to network and be provided with help to polish their personal networking skills.

- Pay close attention to the three types of networkers in your organization: central connectors, brokers, and peripheral people.

- Avoid these roadblocks to building a networking culture: not enough time or skills and lack of focus, commitment, structure, and communications.

Chapter 5: Getting Your People Ready for Open Innovation

- People matter more than ideas when it comes to making innovation of all types happen.

- Your innovation team needs to include people who are good at working in the three-phased discovery-innovation-acceleration (D-I-A) model of innovation put forward by the Radical Innovation Group. Remember that not all people are good at all phases, so you need different people for different phases.

- When looking for people for your team, look for people with these seven skills: (1) intrapreneurial skills, (2) talent for relationship building, (3) strategic influencing, (4) ability to be a quick study (5) balanced optimism, (6) tolerance for uncertainty, and (7) passion.

Chapter 6: What to Consider Before Leaping into a Partnership

Key Chapter Takeaways

- Trust is an essential component of open innovation relationships. As you move toward open innovation, you should begin to look into two questions: (1) What does it take for you to trust others? (2) How do you convince the people in the organizations you want to draw into your open innovation ecosystem to build trust in you and your company and then start forging strong relationships with them?

- Trust is first and foremost established between people and then perhaps between organizations, so it is more important to look at the people side of innovation than the process side.

- The barriers against building trust and strong relationships with stakeholders in your ecosystem may include (1) an internal rather than an external perspective, (2) viewing external partners as someone paid to deliver specific services rather than sources of co-creation and open innovation, (3) being more focused on protecting your own knowledge and intellectual property than opening up and exploring new opportunities, (4) being skeptical about the capabilities of small companies, and (5) being too busy to make it happen.

- Stakeholder management is a critical discipline to master as you try to foster a culture in which trust is a key component and in which resistance to change is minimized so open innovation can thrive.

- You can get an idea of stakeholder management—for both internal and external stakeholders—by thinking in terms of three steps: identification, profiling, and communication.

Key Chapter Takeaways Recap

- Small companies must do their homework to learn as much as possible about the corporation they're about to join forces with, particularly with regard to that organization's history of dealing with smaller companies. This will help you avoid entering into a win-lose relationship in which you get the short end of the deal.

- The more you know about a potential open innovation ally from the get-go, the better prepared you will be understand how your company might be of help to the larger company. This means you will be better prepared to showcase your firm's assets.

- It's easier to start a partnership than it is to end it. But don't be so eager that you don't take time to fully consider whether this is the right partner for you.

- A critical task for smaller companies is to evaluate whether a corporate partner is right for them, given that they have limited resources that a larger company can quickly drain.

- Small companies can provide opportunities for their employees to gain open innovation skills by using innovation intermediaries, teaming up with other businesses in clusters, serving as corporate wild cards, or bringing in consultants to provide training.

Chapter 7: Making It Work

Key Chapter Takeaways

- Important groundwork that should be done at the start of an open innovation relationship includes reaching agreement on goals and defining the desired outcome so everyone is confident that the relationship will produce a win-win that is acceptable to both parties. It also includes establishing boundaries, such as deciding up front how any disagreements will be settled as well as being very clear on what the expectations are as far as the resources that will be devoted to the project from both sides.

- Avoid confidentiality and IP issues in the early stages and facilitate the building of a community.

- Consider town hall events to efficiently make meaningful connections around the globe with potential partners. Use these events to explain the heart and mission of your company, what you're looking for, and what opportunities you have for the attendees to partner with you.

- Put in place a strategy to bring existing partners into your open innovation ecosystem.

Chapter 8: Why Things Go Wrong

Key Chapter Takeaways

- The reasons open innovation partnerships go off track include (1) a lack of capabilities to make innovation happen internally, (2) using idea generation platforms without first defining the criteria for a successful idea, (3) a lack of ability to build trust, and (4) the larger partner trying to manage the process instead of collaborating.

- Open innovation efforts are often killed by corporate antibodies that resist the types of change that being part of an open innovation ecosystem requires.

- Solutions for corporate antibodies include (1) making people backers rather than blockers, (2) staying below the

Key Chapter Takeaways Recap

radar, (3) having a framework and processes in place, and (5) providing high autonomy.

Chapter 9: Innovation Marketplaces: A Major Resource for Open Innovation

Key Chapter Takeaways

- Innovation marketplaces can help small companies meet the challenge of getting employees up to speed on the skills that are required to be part of an open innovation team.
- Small companies can also use the open innovation intermediaries to pose their own challenges and have teams of experts from around the globe work to solve a tough R&D or business problem that may be blocking their success.
- If a small company employee happens to have success with a challenge he or she works on through one of the intermediaries, this will generate enthusiasm for adopting open innovation within the organization.
- InnoCentive, NineSigma, IdeaConnection, TopCoder, and YourEncore each offer a different model; you should explore them all to see which one suits your needs best.

Chapter 10: IPR and Open Innovation

Key Chapter Takeaways

- Concerns about intellectual property issues in open innovation have been decreasing because the companies that are leading the way in open innovation understand that business needs to come before legal issues and they have gained experience in making this happen. The legal departments of these leaders now think offense rather than defense.

- Common mistakes that small companies make when working with large companies in open innovation include (1) thinking they have all the answers to the large company's problems, (2) bringing IP lawyers instead of business-focused lawyers to the table, (3) putting legal issues ahead of business issues, (4) not understanding the economics and low probability of success of product development, and (5) assuming the larger company has bad intentions.

Chapter 11: Using Social Media Tools

Key Chapter Takeaways

- Companies need a multi-target approach to using social media in their open innovation efforts because these efforts lie in three different yet overlapping circles: (1) the innovation community, (2) the innovation ecosystem, and (3) customers and users.
- Social media offers two key aspects that support open innovation: (1) communication of relevant messages and (2) collaboration, including sharing of ideas and solutions.
- LinkedIn, Twitter, and communities built around destination sites created by companies are the three most useful social media tools for open innovation today. But this is a constantly evolving field so you need to stay alert for new developments.
- To help build a status as a preferred partner of choice within your ecosystem, use three elements: (1) the destination site, which is a company-controlled platform/website

that functions as the hub of the community and shows your innovation needs/assets, (2) thought leadership activities through social media channels, and (3) physical events that allow the community members to meet face-to-face.

- The five elements needed to make an open innovation community successful are (1) a genuine need, (2) value, (3) people, (4) communication, and (5) persistence.

APPENDIX A

THE TEN TYPES OF INNOVATION™

Innovation Category	Innovation Type	Description of type	Business example
Finance	1 Business model	How you make money	Dell revolutionized the personal computer business model by collecting money before the consumer's PC was even assembled and shipped (resulting in net positive working capital of seven to eight days).
	2 Networks and alliances	How you join forces with other companies for mutual benefit	Consumer goods company Sara Lee realized that its core competencies were in consumer insight, brand management, marketing, and distribution. Thus it divested itself of a majority of its mfg. operations and formed alliances with mfg. and supply chain partners.

Appendix A

Process			
	3 Enabling process	How you support the company's core processes and workers	Starbucks can deliver its profitable store/coffee experience to customers because it offers better-than-market compensation and employment benefits to its store workers, who are usually part-time, educated, professional, and responsive people.
	4 Core processes	How you create and add value to your offerings	Wal-Mart continues to grow profitably through core process innovations such as real-time inventory management systems, aggressive volume/pricing/delivery contracts with merchandise providers, and systems that give store managers the ability to identify changing buyer behaviors in and respond quickly with new pricing and merchandising configurations.

Offerings			
	5 Product performance	How you design your core offerings	The VW Beetle (in both its original and its newest form) took the market by storm, combining multiple dimensions of product performance.
	6 Product system	How you link and/or provide a platform for multiple products.	Microsoft Office "bundles a variety of specific products (Word, Excel, PowerPoint, etc.) into a system designed to deliver productivity in the workplace.
	7 Service	How you provide value to customers and consumers beyond and around your products	An international flight on any airlines will get you to your intended designation. A flight on Singapore Airlines, however, nearly makes you forget that you are flying at all, with the most attentive, respectful, and pampering pre-flight, in-flight and post-flight services you can imagine.

Appendix A

Delivery	**8** Channel	How you get your offerings to market	Legal problems aside, Martha Stewart has developed such a deep understanding of her customers that she knows just where to be (stores, TV shows, magazines, online, etc.) to drive huge sales volumes from a relatively small set of "home living" educational and product offerings.
	9 Brand	How you communicate your offerings	Absolut conquered the vodka category on the strength of a brilliant "theme and variations" advertising concept, strong bottle and packaging design, and a whiff of Nordic authenticity.
	10 Customer experience	How your customers feel when they interact with your company and its offerings	Harley Davidson has created a worldwide community of millions of customers, many of whom would describe "being a Harley Davidson owner" as a part of how they fundamentally see, think, and feel about themselves.

Copyright © Doblin. All rights reserved. Used with permission.

Made in the USA
Charleston, SC
30 October 2011